Advances in

ECOLOGICAL RESEARCH

VOLUME 28

Advances in

ECOLOGICAL RESEARCH

Edited by

A.H. FITTER

Department of Biology, University of York, UK

D. RAFFAELLI

Department of Zoology, University of Aberdeen, UK

VOLUME 28

ACADEMIC PRESS

San Diego London
Boston New York
Sydney Tokyo Toronto

This book is printed on acid-free paper.

Academic Press, Inc.
525 B Street, Suite 1900, San Diego, California 92101–4495, USA
http://www.apnet.com

Academic Press Limited
24–28 Oval Road, London NW1 7DX, UK
http://www.hbuk.co.uk/ap/

ISBN 0–12–013928–6

A catalogue record for this book is available from the British Library

Typeset by Saxon Graphics Ltd, Derby
Printed in Great Britain by MPG Books Limited, Bodmin, Cornwall

99 00 01 02 03 04 MP 9 8 7 6 5 4 3 2 1

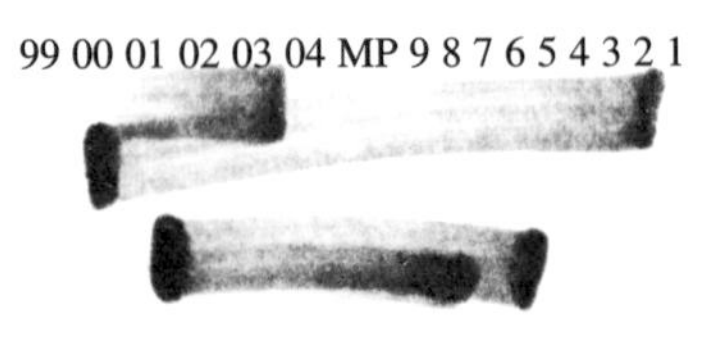

Contributors to Volume 28

S.R. BALCOMBE, *Co-operative Research Centre for Freshwater Ecology, Department of Environmental Management and Ecology, La Trobe University, P.O. Box 821, Wodonga, Vic. 3689, Australia.*

T.M. BLACKBURN, *NERC Centre for Population Biology, Imperial College at Silwood Park, Ascot, Berkshire SL5 7PY, UK.*

S.P. BLYTHE, *Department of Statistics and Modelling Science, University of Strathclyde, Glasgow G1 1XH, UK.*

R.M. CALLAWAY, *Division of Biological Sciences, University of Montana, Missoula, MT 59812, USA.*

G.P. CLOSS, *Department of Zoology, University of Otago, P.O. Box 56, Dunedin, New Zealand.*

K.J. GASTON, *Department of Animal and Plant Sciences, University of Sheffield, Sheffield S10 2TN, UK.*

S.J. GOETZ, *Department of Geography, Laboratory for Global Remote Sensing Studies, University of Maryland, College Park, MD 20742, USA.*

W.S.C. GURNEY, *Department of Statistics and Modelling Science, University of Strathclyde, Glasgow G1 1XH, UK.*

S.P. LONG, *Departments of Crop Science and Plant Biology, University of Illinois, 379 Edward R. Madigan Laboratories, 1201 W. Gregory Drive, Urbana, IL61801, USA.*

A.R. MCLEOD, *Centre for the Study of Environmental Change and Sustainability, John Muir Building, The Kings Buildings, Mayfield Road, Edinburgh EH9 3JL, Scotland, UK.*

B.E. MAHALL, *Department of Ecology, Evolution and Marine Biology, University of California, Santa Barbara, CA 93106, USA.*

S.D. PRINCE, *Department of Geography, Laboratory for Global Remote Sensing Studies, University of Maryland, College Park, MD 20742, USA.*

H.J. SCHENK, *National Center for Ecological Analysis and Synthesis, 735 State Street, Suite 300, Santa Barbara, CA 93101-3351, USA.*

M.J. SHIRLEY, *Co-operative Research Centre for Freshwater Ecology, Department of Biological Sciences, Monash University, Clayton, Vic. 3168, Australia. (Current address: Environmental Sciences, Water ECOscience, Private Bag 1, Mount Waverley, Vic. 3149, Australia.)*

T.K. STOKES, *CEFAS, Lowestoft Laboratory, Pakefield Road, Lowestoft NR33 0HT, UK.*

Preface

This volume contains six reviews which address modelling, experimental and empirical approaches to pure and applied ecological issues at scales ranging from local to landscape to global. The articles cover macroecology, food webs, population dynamics and biology, large-scale experiments, atmospheric carbon and remote sensing. Notwithstanding this diversity, the links between these reviews and their points of contact are many, and there is a general theme of the need to match modelling with experiments, and patterns with processes.

One of the most immediate and technically challenging issues facing environmental scientists today is predicting the effects of enhanced levels of carbon dioxide on natural ecosystems. Many of the experiments which have been carried out to examine such effects have traditionally been small-scale, with all the limitations and potential artefacts associated with such designs. These are largely overcome by the large-scale FACE (Free-Air Carbon dioxide Enrichment) approach. McLeod and Long describe the application of this approach to the performance of several vegetation types (cotton, wheat, ryegrass, clover and loblolly pine trees). The results of the experiments carried out to date are exciting and sometimes contrary to expectations. The FACE experimental approach has great potential for understanding the ecological impacts of climate change.

The atmospheric carbon theme is picked up by Goetz and Prince's review. Tackling the issue of carbon exchange and storage at a global scale demands data at the appropriate scale and resolution as well as models which address all the processes concerned. These authors provide a compelling argument that large area primary production models which are driven by light absorption need to include quantitative terms reflecting the actual respiratory costs of maintenance and synthesis if they are to make the best use of remotely sense information.

Analyses of well-documented, real food webs have shown that several properties thought to enhance the stability of ecological systems are hard to find. In particular, omnivores are more common than suspected, food chains are relatively long and there is a high number of feeding links per species. Clearly, many predators are generalists, and Closs and his colleagues show how the stabilising (or destabilising) influence of these predators is scale-dependent. At small spatio-temporal scales they cause local extinction of prey, but when viewed at the larger scale, they can be a stabilising influence, especially if the prey have local refugia. The relative importance of top-down and bottom-up regulation in food webs may be strongly linked to such processes.

The defence of territorial space by animals is well-documented and its function for avoiding the costs of scramble competition generally well-accepted. The question of territoriality in plants, how such territoriality might be expressed and its precise function, has long intrigued plant population ecologists. Here, Schenk and his colleagues address these issues with respect to the spatial segregation of root systems, showing that when space is physically restricted, plant growth declines, even when other resources are abundant. Plants also appear to compete for space independently of other limiting resources. Active root segregation and the defence of below-ground space show remarkable parallels to equivalent mechanisms used by territorial animals.

An often forgotten feature of the way scientists practice their art is that it is only possible to prove a particular hypothesis to be false. Yet much of what scientists set out to find concerns the discovery of what is true. The Popperian paradigm of hypothesis generation followed by refutation is rigorous but, given the often extensive range of plausible alternative hypotheses available, most experienced scientists are not content to simply compile long lists of incorrect explanations. Instead, they use their experience and intuition to form an idea as to how their bit of the world works. Basically, they painstakingly build up a jigsaw picture of processes, mechanisms and explanations which are consistent with the data available. Gurney and his colleagues suggest that this process is similar to the concept of reasonable doubt embraced by those in the legal professions. They illustrate this process by re-analysing a classic data set on blowflies, first described by Nicholson. Their analyses have general implications for the design of experimental and observational programmes.

Finally, Blackburn and Gaston provide a critical and incisive analysis of the mechanisms which have been proposed to account for empirical relationships between animal abundance and body size. This area of macroecology has seen a huge expansion of interest (and publications) over the last few years as more data sets become available and are analysed in different ways. Notwithstanding criticisms about quality of the data sets used, there has been a tendency to attribute the observed relationships to constraints placed on abundance by the energy available within body size classes. However, there are at least five other competing hypotheses to explain these patterns and these are critically assessed in this review. In particular the energy constraint hypothesis is shown to be logically flawed. They offer an alternative explanatory model based on the distribution of biomass between species and a constraint on minimum viable abundance.

Dave Raffaelli
Alastair Fitter

Contents

Generalist Predators, Interaction Strength and Food-web Stability

G.P. CLOSS, S.R. BALCOME AND M.J. SHIRLEY

Delays, Demography and Cycles: A Forensic Study

W.S.C. GURNEY, S.P. BLYTHE AND T.K. STOKES

Spatial Root Segregation: Are Plants Territorial?

H.J. SCHENK, R.M. CALLAWAY AND B.E. MAHALL

The Relationship between Animal Abundance and Body Size: A Review of the Mechanisms

T.M. BLACKBURN AND K.J. GASTON

Free-air Carbon Dioxide Enrichment (FACE) in Global Change Research: A Review

A.R. McLEOD AND S.P. LONG

I. SUMMARY

There have been many experimental studies to evaluate the response of vegetation to the effect of increases in the partial pressure of carbon dioxide in the atmosphere (pCO_2) that are expected to occur during the next century. This knowledge is important for the future protection of food supplies, for understanding changes in natural ecosystems and for quantifying the role of terrestrial plants in regulating the rate of change of pCO_2 and resulting

ADVANCES IN ECOLOGICAL RESEARCH VOL. 28
ISBN 0-12-013928-6

changes in the global climate. Most of our knowledge about these effects has derived from experimental studies that have used open-top or closed chambers. These methods are subject to "chamber effects" caused by differences in energy balance and water relations that may significantly modify the response of vegetation to elevated pCO_2. The small plot sizes imposed by these techniques add other limitations both to interpretation of results and scope of investigations. Free-air carbon dioxide enrichment (FACE) provides an experimental technique for studying the effects of elevated pCO_2 on vegetation and other ecosystem components in large unenclosed plots (>20 m diameter). FACE avoids many modifications to the microclimate imposed by chamber methods and therefore provides some of the most reliable estimates of plant response to elevated pCO_2. Control of pCO_2 in large-scale FACE experiments has now been developed to an extent where performance is similar to that achieved with sophisticated closed-chamber facilities. Experience has shown that, when FACE facilities are fully utilised, the cost per unit of usable ground area enriched with CO_2 is significantly lower than alternative methods. The large scale of FACE plots can support a range of integrated studies on the same material, thereby achieving a more complete analysis than has been possible with other methods of elevating pCO_2. This review considers the technical aspects of FACE methodology, outlines the major FACE experiments and summarises the advances in understanding of pCO_2 effects on ecosystems that it has allowed.

Published data on large-scale FACE experiments with adequate plot replication are limited to experiments on four crop/vegetation types at three locations. FACE has been used for experiments on two crops, cotton and wheat, at Maricopa, Arizona, and on grassland species, principally ryegrass and clover, at Eschikon, Switzerland. The method has also been adapted for the first study of mature forest trees, loblolly pine at Duke Forest, North Carolina. A number of other large-scale FACE experiments are in progress and the method has been adapted for use in much smaller experimental plots.

The results of the major FACE experiments represent important advances over understanding obtained from previous pCO_2 treatment methods. Most significant in terms of the global climate and atmosphere system are the clear observations with cotton and wheat crops that elevated pCO_2 increases the ratio of sensible : latent heat transfer and causes daytime warming of the surface vegetation. This results from decreased water use and loss, and has been evident at a range of scales. The scale of FACE plots has allowed quantitative and detailed studies of the dynamics of below-ground production and C accumulation in a range of systems, and all have shown surprisingly large increases. Of particular note are the increases observed in grassland grown with low N, where there was no response of the above-ground biomass, but an increased rate of turnover of leaves and input of surface litter. FACE has allowed cultivation of crops at a scale appropriate to agronomic trials and shown statistically significant increases in the yields of wheat, cotton and pasture crops, although some of these increases are less than suggested by chamber experiments. Against expectations, the FACE experiments at Maricopa have shown a greater relative increase in yield in crops grown under water shortage than in water-sufficient crops. Acclimatory loss of photosynthetic capacity has been widely anticipated to offset the increase in photosynthesis that follows initial transfer of vegetation to elevated pCO_2. None of the FACE experiments provides any evidence of such a loss; however, changes which will allow a re-optimisation of N distribution within plants have been reported.

FACE methods have now been demonstrated to be feasible and effective within a range of crops and vegetation types. The information from past experiments has greatly improved our understanding of the impacts of global atmospheric change on terrestrial ecosystems.

II. INTRODUCTION

Free-air carbon dioxide enrichment (FACE)[1] systems (Hendrey, 1993) provide a large-scale outdoor experimental technique for studying the effects of elevated CO_2 concentration on vegetation and many other ecosystem components. FACE experiments utilise a distribution system of pipes or plenums surrounding an unenclosed experimental plot and release a mixture of CO_2 in air to elevate the partial pressure of CO_2 (pCO_2) within and around the plant canopy. The gas emitted is further diluted close to the release points, primarily by horizontal advection and secondarily by vertical turbulent eddy diffusion. Enrichment of air by emission jets and continuous mixing by turbulence reduces temporal and spatial fluctuations in pCO_2 (Allen *et al.*, 1992). A personal computer-based control system is used to determine the rate and position of CO_2 release from gas sources. A rapid updating and correction of CO_2 release rate with a control algorithm based on rapid, direct measurements of wind speed, wind direction and pCO_2 at the centre of a plot provides a close control of the exposure concentration. A FACE system of the Brookhaven National Laboratory (BNL), USA, is shown in Figure 1 installed in a mature forest canopy. FACE-type systems have been designed for studies of elevated pCO_2 in crops, forests and natural vegetation (Table 1) and have also been used to study atmospheric pollutant effects (McLeod, 1993).

Why were FACE systems developed? Our understanding of how crop and natural ecosystems will respond to the anticipated increase in pCO_2 over the next century is based largely on laboratory-based controlled environment and field-chamber studies. The majority of field studies have been based on the use of open-top chambers (OTCs). Despite the fact that the top, or the

---

[1]Symbols and abbreviations are listed in the Appendix (p. 55).

Fig. 1. View of the prototype Forest FACE facility (FACTS-I) at Duke Forest, North Carolina, USA. (Photograph courtesy of G.R. Hendrey, Brookhaven National Laboratory, USA.)

larger portion of the top, of an OTC is open to the atmosphere, there are still important differences between the environment within the best-engineered OTCs and the environment surrounding the chamber. As a result, chamber effects, when carefully monitored, can be as great as the effect of doubling pCO_2 within the chamber (e.g. Knapp *et al.*, 1994; Day *et al.*, 1996; Drake, 1996). Whitehead *et al.* (1995) evaluated the performance of large OTCs of the design of Heagle *et al.* (1989) and compared microclimatic conditions with those outside. When photon flux was 1600 μmol m^{-2} s^{-1}, air temperature was 4.3°C higher than outside and humidity 0.8 Pa lower. In cloudy conditions the mean transmittance of solar irradiance into the chambers was 81% and on clear days this decreased from 80% to 74% with increasing solar zenith angle. The ratio of diffuse to total solar irradiance in the chambers was 13% and 21% greater than that outside for cloudy and clear conditions, respectively. Transmittance of visible solar irradiance (400–700 nm) through the plastic wall material had decreased by 7% after 1 year of exposure at the site. Other obvious effects of an OTC are that wind is removed, preventing wind damage and dispersal of pathogens and pests, rainfall interception is decreased and plant–atmosphere coupling is altered. Most materials used in the construction of OTC walls do not transmit wavelengths of ultraviolet-B radiation (280–315 nm). These factors may contribute to a varying extent to

Table 1
Published details of FACE experiments with replicated plots (>100 m^2) and in operation on 1 January 1998

Vegetation	Location	Year	pCO_2 (Pa) (± 24 h)	Other Treatments	Replicates	Ring Diameter (m)	Reference
Cotton	Farm at Maricopa, AZ, USA	1989	55	None	4E + 4C	23	Lewin *et al.* (1994)
Cotton		1990/1	55	2 water (split-plot)	4E + 4C	25	Mauney *et al.* (1994)
Wheat		1993/4	55 (24 h)	2 water (split-plot)	4E + 4C	25	Kimball *et al.* (1995)
Wheat		1996/7	Amb. + 20 (24 h)	2 N (split-plot)	4E + 4B + 2C	25	Pinter *et al.* (1996)
Perennial ryegrass	ETH, Eschikon, Switzerland	1993–1998	60, s	2 N × 2 cutting freqs.	3E + 3C	18	Blum (1993)
White clover plus other pasture spp.				L.p, T.r and L.p. × T.r. plus other expts.			Hebeisen *et al.* (1997b)
Loblolly pine (>12 m height)	FACTS-I Duke Forest, NC, USA	1996 onwards	Amb. + 20 (24 h)	None	3E + 3B	30	Hendrey *et al.* (1998)
Aspen Sugar maple Paper birch	FACTS-II Rhinelander, WI, USA,	1998 onwards	55	60 ppm-hr O_3 with/without CO_2	3E, 3B, 3O, 3E with O	30	Isebrands *et al.* (1998)
Desert scrub and C_4/C_3 grasses	Mojave Desert, NV, USA	1997 onwards	55	None	3E + 3B + 3C	23	D. Jordan (personal communication)
Tall grass prairie	Cedar Creek, Bethel, MN, USA	1997 onwards	55, s	4 levels of species × 2 N in 4 m^2 sub-plots	3E + 3C	24	G. Hendrey (personal communication)

pCO_2 column: 24 h = day and night fumigation (otherwise daytime only); s = growing season fumigation only; Amb. = site ambient partial pressure.
Replicates column: E = elevated pCO_2; B = blowers, without elevated pCO_2; C = control without blowers; O = elevated pO_3.
Other treatments column: L.p = *Lolium perenne*; T.r = *Trifolium repens.*

changes reported in the growth form of vegetation inside an OTC (Murray *et al.*, 1996; Tissue *et al.*, 1996). The effect of elevated pCO_2 on ecosystems and crops are therefore based on OTC measurements made in an aerial microclimate that may be duller, warmer and drier than the norm. Since FACE avoids these modifications of microclimate, one role of FACE experiments has been to test the applicability of observations in OTC experiments and to provide a more exact representation of how elevation of pCO_2 in isolation will affect ecosystems.

A further major advantage presented by the FACE technique is the larger scale of plot that may be used compared to that in OTCs and other replicated enclosures. This minimises the influence of edge effects that have a greater influence in field chambers where the edge/area ratio is much larger. The large area available in FACE experiments is of particular relevance in uniform vegetation, such as arable crops, plantation forests and temperate/boreal forests. In other situations, where vegetation is very heterogeneous, with only a few individuals of each species per plot, the scale of large FACE plots may still be inadequate. Large FACE plots allow simultaneous analysis and sampling by many different research groups and interaction between them. For example, in early FACE experiments (detailed in section III), studies have included leaf and canopy gas exchange, biochemical and molecular analysis of photosynthesis, leaf area and canopy development, above- and belowground biomass accumulation, losses, and final yield. Water studies have included simultaneous measurement of leaf and canopy transpiration, canopy energy balance, stem water flow, soil moisture and extraction. A complete analysis of this type is possible only with the scale of the plots that FACE can support. As such it provides the most complete data sets for validating process-based models (Grant *et al.*, 1995; Grossman *et al.*, 1995).

Allowing access of many groups to FACE plots might in itself be a potential problem. Trampling would alter the canopy form and create the edge effects noted as a serious potential limitation of OTCs. The early FACE experiments have now demonstrated that use of defined permanent walkways, strict rules and pre-defined sampling plans, and on-site technical supervision, will restrict disturbance to levels similar to those in a commercial crop (Mauney *et al.*, 1994).

This review aims to:

(1) provide a background to FACE methodology;
(2) outline the major FACE experiments and to summarise their results;
(3) consider the advantages and disadvantages of FACE in a biological context relative to other enrichment techniques;
(4) determine the value of the scientific findings from FACE experiments to our understanding of global atmospheric change.

III. FACE EXPERIMENTS AND THEIR DEVELOPMENT

A. How FACE Systems Work

In 1986 the BNL began the development of a prototype FACE following a number of earlier trials and gas dispersion simulations described by Allen (1992). The prototype trials led to the establishment of a full field experiment at the Maricopa Agricultural Center (MAC) of the University of Arizona. This design consisted of four, independently controlled, circular, FACE arrays and four control plots. Enrichment was continuous during daylight hours to a set point of 55 Pa. Full details of the design, operation and performance of the system are given by Lewin *et al.* (1994) and by Nagy *et al.* (1994).

Each plot was surrounded by a 25 m diameter toroidal plenum constructed of 0.3 m internal diameter pipe with 2.5m high vertical pipes fitted with individual control valves spaced at 2m (see Figure 2). Air enriched with CO_2 is blown into the plenum and exits through tri-directional jets in the vertical pipes near the top of the canopy. Wind direction, wind velocity and CO_2 concentration are measured at the centre of each plot and this information is used by a computer-controlled system to adjust CO_2 flow rate, regulated by a mass-flow control valve and thereby maintain the target pCO_2 of 55 Pa. Only pipes on the upwind side of the plots release enriched air unless wind velocity is less than 0.4 m s^{-1} when it is released from alternate vertical release pipes. This basic FACE system has been utilised with some variations and technical developments in many subsequent experiments including studies of cotton, wheat, grassland and desert ecosystems, and forest trees (Table 1).

B. FACE in Crops and Grassland

1. FACE Experiments on Cotton at Maricopa, AZ, USA

In 1989 a full field experiment comprising four FACE plots of 25 m diameter, each blocked with four control plots, was set up to study effects of elevated pCO_2 on cotton (*Gossypium hirsutum* L. cv. Deltapine 77) over 3 years: 1989, 1990 and 1991. The experiments were conducted in a large field at the experimental research farm of the University of Arizona at Maricopa, about 50 km south of the city of Phoenix, on an extensive fertile and flat alluvial plain. Each crop was sown over the entire field in mid-April following standard agricultural practice for the region and then the FACE dispersion pipes were put into place around each plot before emergence. An elevated pCO_2 treatment of 55 Pa was imposed continuously from about the day of emergence until the crop was ready for harvest in mid-September (Mauney *et al.*, 1994; Pinter *et al.*, 1996). Daytime temperature averaged over 30°C. From the 1990 growing season onwards each plot was divided in half and a sub-soil drip irrigation used to provide a fully irrigated and a drought treatment. In the

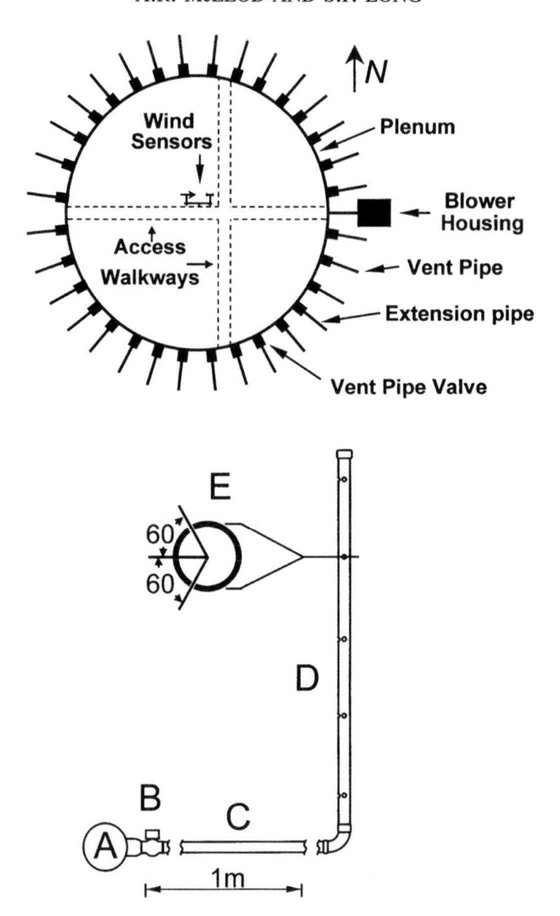

Fig. 2. Diagram of a BNL FACE system as used at Maricopa, Arizona. Major components of a single array and a detailed view of a single vent pipe showing: (A) distribution plenum, (B) quarter-turn ball valve, (C) extension pipe, (D) vertical vent pipe showing five sets of emission ports, and (E) enlarged view of a cross-section of the vent pipe through a set of emission ports showing the orientation of the adjacent ports (redrawn from Lewin *et al.*, 1994).

fully irrigated treatment, water was replaced according to *ET* (potential evapotranspiration). In the drought treatment, water was replaced to 0.75 of *ET* in 1990 and 0.67 of *ET* in 1991 (Mauney *et al.*, 1994). Results obtained were analysed as a split-plot, with $p\mathrm{CO_2}$ as the split-plot factor within each block. The four blocks were treated as the individual replicates in assessing the statistical significance of CO_2 effects on individual variables. This avoided any pseudo-replication in the analysis. The facility was used by 47 investigators from 23 research organisations during the three experiments and produced a large amount of information on the growth and physiological responses of cotton to elevated pCO_2 at ample and limiting water supplies. These results are reported in special volumes (Hendrey, 1993; Dugas and Pinter, 1994). The results suggested that, averaged over the 3 years of study, cotton yields were increased by 40% by pCO_2 elevated to 55 Pa and there was no significant increase in water use.

2. *FACE Experiments on Wheat at Maricopa, AZ, USA*

After the FACE-cotton experiments of 1989–1991, the FACE facility at the MAC was used from 1993 for experiments on wheat crops (*Triticum aestivum* L. cv. Yehora Rojo) (Kimball *et al.*, 1995; Pinter *et al.*, 1996). The wheat crops were planted in late December 1992 and harvested in May. Daytime temperatures averaged about 20°C for the wheat crop growth period. Treatments with ample and limiting levels of water supply were nested within the plots during the first two experiments (Table 1). As for cotton, water was replaced in the fully irrigated treatment according to *ET*. In the drought treatment, water was replaced at 0.5 of *ET* for the wheat crops in 1993 and 1994 (Pinter *et al.*, 1996). Over 50 scientists from eight countries participated in the FACE-wheat experiments. Studies included: plant growth (leaf area, height, biomass, apical and morphological development), canopy temperature, reflectance, chlorophyll, light-use efficiency, energy balance, evapotranspiration, soil and plant analyses, soil water content, sap flow, leaf and canopy photosynthesis, respiration, stomatal conductance, leaf water potential, carbohydrate dynamics, photosynthetic proteins, gene transcript levels, antioxidants, stomatal density and anatomy, digestibility, decomposition, grain quality, soil CO_2 fluxes and changes in soil C storage. The results from these early wheat experiments suggest that the responses of wheat were different from those of cotton in the elevated pCO_2. During the cool early season, there was little response to CO_2 but later in the season, as temperature rose, the elevated pCO_2 plants grew about 20% more than ambient controls with the mean tiller number increased by one. The plants in elevated pCO_2 plots were about 0.6°C warmer than controls both day and night in the well-watered plots where water use was decreased by 5–12%. Elevated pCO_2 and water stress interacted to cause an 8–12% yield increase in well-watered plots but an increase of 21–25% in the droughted plots. Subsequent experiments with wheat from December 1995 onwards replaced the water stress treatments with ample and limiting levels of soil nitrogen (Table 1). Grain yields were increased by about 15% in the 1995–1996 experiment under elevated pCO_2 and ample N with a 12% increase at low N. The low N treatment reduced yield by 20% at both levels of pCO_2 and there was no change in water use. The FACE-wheat experiment with ample and limiting N was repeated during 1996–1997.

3. FACE Experiments on Grassland and Desert Ecosystems

During 1992, a FACE system based on the earlier BNL design was constructed on farmland adjacent to the experimental station of the ETH (Swiss Federal Institute of Technology) at Eschikon about 20 km north of Zurich, Switzerland, for studies in a grassland ecosystem (Blum, 1993; Hebeisen *et al.*, 1997b). The system differed from previous designs in that the gas supply plenum was buried underground and the vertical gas release pipes emerged 2 m inside the plenum circle, thus providing potential for expansion of plot size without the need to modify plenum size. Three FACE plots of 18 m diameter were established in August 1992, each blocked with three control plots according to cropping pre-history, and have now been operated over the course of 5 years (Jongen *et al.*, 1995; Hebeisen *et al.*, 1997b). Perennial herbage crops were sown to a common design in all six plots. The area within each plot is partitioned into a series of sub-plots separated by access paths, allowing a series of experiments to be undertaken (Pearson *et al.*, 1995; Ineson *et al.*, 1996; Hebeisen *et al.*, 1997b; Nitschelm *et al.*, 1997; Bucher *et al.*, 1998; Luscher *et al.*, 1998). The central portion of each plot is occupied by sub-plots of pasture cultivars of perennial ryegrass (*Lolium perenne* L. cv. Bastion) and white clover (*Trifolium repens* L. cv. Milkanova). Twelve sub-plots (2.8 m × 1.9 m) were used in a factorial design with sub-plots of the two species as monocultures and a mixture with equal amounts of the two, provided by halving the sowing rates used in the monocultures. Cultivation followed standard practice for these crops at the location. Two of the four plots of each monoculture and of the bispecies mixture received a high N application (56 g m^{-2}) and two a low N application (14 g m^{-2}). The crops were cut to 5 cm above the ground surface at each harvest. Within each N treatment, one sub-plot was cut seven or eight times in the year (intensive) the other four times (extensive) (Hebeisen *et al.*, 1997a,b).

This sub-division of the space within a large FACE plot would appear to lose the advantage of scale gained in the Maricopa experiments with arable crops. However, since these were short herbage crops, the plot size required to avoid edge effects was sufficiently small to allow the use of these small sub-plots and still harvest from areas that would not be subject to edge effects. The approach of using small experimental plots within a large FACE rings is discussed in section V.E.

The FACE system at Eschikon was operated with a target pCO_2 concentration of 60 Pa imposed from May 1993. Treatment occurred during daylight hours and throughout the growing season when air temperature at 2 m equalled or exceeded 5°C in March until it fell to this threshold in November (Hebeisen *et al.*, 1997b). Results obtained were analysed as a split-plot, with pCO_2 as the split-plot factor within each block and effects of nitrogen or cutting frequency analysed as split-split-plot designs. Statistical significance has been assessed by treating the three blocks as the individual replicates, avoiding pseudo-replication in the analysis (Fischer *et al.*, 1997).

The BNL design is also being used to construct FACE facilities with replicated plots for study of a desert ecosystem in Nevada, USA (D. Jordan, personal communication), and a prairie grassland in Minnesota, USA (G. Hendrey, personal communication) (Table 1).

4. FACE Experiments in Forests

A study was undertaken in 1993–1994 by BNL to develop a FACE system for control of pCO_2 in a tall forest canopy (Hendrey *et al.*, 1994, 1998). Two models of FACE system were built at the Duke Forest, NC, USA, termed FFP-1 (Forest FACE prototype 1) and FFP-2. The FFP-1 design, covering a 30 m diameter plot dominated by 8–10 m tall loblolly pine *(Pinus taeda)*, was operated for 2 months in 1993. This identified possible design improvements that were incorporated in the FFP-2 design built in Spring 1994. The FFP-2 system covered a 34 m diameter plot and was tested on 38 12-hour days in summer 1994 with a target pCO_2 of 55 Pa. The daily mean varied by 1 Pa with an average value of 55 Pa (3% lower than the set point). Further details of performance are given by Hendrey *et al.* (1994, 1998). The Forest FACE design of BNL has been further developed for use in a study entitled "Forest-atmosphere carbon transfer and storage – I (FACTS-I) (G. Hendrey, personal communication). A system of three FACE plots of 30 m diameter, each blocked with three control plots were erected in 1996 in a 14-year-old 32 ha block of loblolly pine (*Pinus taeda* L.) forest of 12 m height. The elevated pCO_2 treatment has been imposed continuously day and night from August 1996, except when temperatures dropped below 5°C for more than 1 hour. Rather than the fixed set point for the elevated treatment used in the preceding experiments, the elevated pCO_2 treatment in this experiment was set at 20 Pa above ambient. On average this resulted in an elevated pCO_2 treatment of 57.4 Pa in 1997 (Ellsworth, 1998).

Further FACE experiments in forest systems are also in progress. A system using the BNL design, to be known as FACTS-II, is under development for studies of increased pO_3 and pCO_2 in pure and mixed stands of aspen (*Populus tremuloides*), sugar maple (*Acer saccharum*) and paper birch (*Betula papyifera*) at Rhinelander, Wisconsin (Isebrands *et al.*, 1998). A system for studies in a deciduous forest comprising sweetgum (*Liquidambar styraciflua*), is also under construction in the Oak Ridge National Environmental Research Park, TN, USA (R. Norby, personal communication).

5. Other FACE Systems

The apparent advantages of large-scale FACE plots are counteracted by the large amount of CO_2 gas required and its associated cost. Consequently, there have been attempts to design and operate small FACE systems. Maintaining effective spatial and temporal control of pCO_2 in small plots presents a number of additional difficulties. In a small plot there is a reduced distance between the gas sources and the target vegetation for adequate mixing to occur. Consequently, the potential for gradients in pCO_2 across the plot and the presence of large and rapid fluctuations is increased. Control of pCO_2 will also be difficult because the smaller plot size in relation to the scale of turbulence above vegetation will make pCO_2 more susceptible to the effect of turbulent eddies from above the canopy.

Theoretical improvement in the spatial distribution of pCO_2 could be achieved by releasing sufficient gas at the desired pCO_2 to avoid the need for dilution. However, this would need to displace the bulk of the airflow close to the canopy that would be difficult even at mean wind velocities. It would also remove the advantage of large-scale FACE of having realistic wind velocities and canopy-atmosphere coupling. An alternative approach is to improve the rate of mixing after release of gas from vent pipes. In order to evaluate a range of potential designs for smaller FACE systems, a numerical simulation model was developed by Xu *et al.* (1997) and applied to a full-size FACE design. Computational fluid dynamics modelling, wind-tunnel studies and field trials were also used to develop a new method for improving FACE system performance on small plot areas, called enhanced local mixing (ELM) (Walklate *et al.*, 1996). The ELM system uses a special array of baffle elements to enhance vertical and lateral mixing of air jets and so provides a theoretical improvement in gas-use efficiency of 40% when used in 5 m diameter plots (Walklate *et al.*, 1996). The ELM system consequently improves gas mixing and the potential to use FACE gas-release pipes surrounding small plots. Enhancement of gas mixing using baffles also has the potential for use in studies of other atmospheric gases such as O_3, where vegetation may be sensitive to very short periods of high concentration.

The conventional FACE design of BNL has also been scaled down to an 8 m diameter plot termed mid-FACE by Miglietta *et al.* (1997) and used for field experiments on potato (*Solanum tuberosum* L.). Even smaller designs termed miniFACE have been reported for 1 m square plots by Miglietta *et al.* (1996) and for 10 cm diameter plots by Spring *et al.* (1996). These very small miniFACE designs (< 1 m^2) have a continuous air flow, producing leaf movement which dominates the turbulence impacting the vegetation under most conditions. This modifies plant–atmosphere coupling and may influence responses to elevated pCO_2 (as discussed later in section V.B.1) and thus eliminates one of the main advantages of large-scale FACE.

IV. TECHNICAL PERFORMANCE OF FACE SYSTEMS

The technical performance of FACE systems is of great importance for interpretation of biological responses to the CO_2 treatments. There are currently only three fully operational FACE facilities, all designed and assembled by BNL, for which there is detailed information on technical performance. The first is that operated at MAC, AZ, USA, the second is that operated by ETH, Zurich, Switzerland, and the third is at Duke Forest, NC, USA. Descriptions of technical performance in other FACE systems have been limited. In particular, information on spatial and temporal performance is often constrained by the use of long-averaging times for mean concentration values and data are presented only for short periods which are not representative of the full range of operating conditions (Reece *et al.*, 1995; Miglietta *et al.*, 1996). Fluctuations in pCO_2 in both open-top chambers and in FACE systems result from the CO_2 control system itself, or from wind mixing ambient air with the CO_2-enriched air. The measured and reported variations in pCO_2 and control system performance are dependent upon the response characteristics of the CO_2 monitor, the length and diameter of sample lines, the sample flow rate and the use of any vessel as a volumetric buffer to limit pCO_2 variation. Many sampling systems would not permit the resolution of fast fluctuations of pCO_2 within a FACE plot or chamber. Farrow *et al.* (1992) undertook a theoretical analysis of the BNL-FACE sampling system to estimate the effect of each component on system performance. The possible biological significance of short-term fluctuations in pCO_2 is discussed below. Published information on system performance and pCO_2 should always be considered in relation to the characteristics of the sampling system and the averaging time used in calculations.

A. Spatial Distribution of Gas Concentration

Knowledge of the spatial distribution of pCO_2 within a FACE plot is important in order to define the usable area of the plot which experiences an exposure within defined limits of the concentration at the central sampling point typically used for measurement and control. On large-scale systems, e.g. BNL-FACE, it has been observed, as modelling predicts, that short-term concentrations on an upwind edge are significantly higher than at the centre and that downwind edge concentrations are only slightly lower than at the centre (Nagy *et al.*, 1994). Over longer periods of time, changes in wind direction tend to average out this effect to produce a bowl-shaped distribution with regions closer to the emitters experiencing slightly higher concentrations than the centre. Ideally, the distribution of wind directions and associated atmospheric stability and turbulence should be symmetrical. However, in practice, dominant wind directions may be apparent in data on the spatial distribution of pCO_2 (Nagy *et al.*, 1992). FACE plots should also be sited away from large obstacles that could produce turbulence and thus modify FACE performance.

The usable area within the plots of the FACE-cotton experiments was established to within 2 m of the gas outlets (Mauney *et al.*, 1992, 1994) based upon observations that the crop growth response was uniform within this area. This definition of a "usable" area for biological research was supported by measurements collected over a continuous 25-day period (Nagy *et al.*, 1994). The long-term average pCO_2 within the 20 m diameter usable area was 56 Pa. Examination of the spatial variation of mean concentration, partly influenced by wind directions, revealed a standard deviation of about 8 Pa near the plot centre and about 12 Pa in zones towards the edge. Further evidence that pCO_2 did not vary substantially within the plots was provided by measurements of $\delta^{13}C$ incorporation into plant tissue which suggested that mean pCO_2 was held at 55 Pa ± 6% (Leavitt *et al.*, 1994). However, in an earlier study (Nagy *et al.*, 1992), a smaller usable area or "sweet spot" was defined as the zone in which concentrations were ± 20% of the target for > 80% of the time. This was 12 m in diameter within a 23 m FACE plot. The usable area in the Maricopa FACE experiments was based on modelling, and the observation that plant growth and plant $\delta^{13}C$ appears to integrate any short-term fluctuation in spatial distribution of pCO_2. This assumption may not be appropriate if FACE-type systems are used for studies with other atmospheric gases such as the pollutants SO_2 and O_3.

In circular release systems, attention must be paid to the concentrations produced at the edges, where sources may effectively be spaced closer together along an axis perpendicular to the wind or totally aligned along the wind-direction axis for a square release system. The BNL FACE design was modified by reducing the number of gas outlets that open at the edges of the crosswind axis ("feathering": Nagy *et al.*, 1992). However, later information suggests that this effect may overcompensate to produce lower concentrations at the edges of the crosswind axis (Nagy *et al.*, 1995). Information of this type is lacking for other systems and is likely to be particularly important for small FACE designs.

The development of a Forest FACE system has presented an even greater technical challenge to achieve uniformity of spatial and temporal concentration patterns because of the non-uniform structure of the wind profile within a mature tree canopy (Hendrey *et al.*, 1998). The usable area of vegetation within the FACE-cotton experiments was defined up to 2 m from the gas outlets (Mauney *et al.*, 1992, 1994). However, if this design were used for an experiment with phytotoxic gases such as O_3 in combination with CO_2, the peak concentrations close to the outlets would result in unrealistic visible injury to the vegetation. If a larger buffer zone of unusable vegetation is adopted, requiring larger diameter arrays of release pipes, the cost of CO_2 becomes very much greater. Possible alternative approaches include a larger diameter array for release of phytotoxic gases outside a smaller diameter array for release of CO_2, or the possible use of mixing baffles as described earlier. This potential problem highlights the fact that large variations in pCO_2 do exist close to pipe outlets in full-size FACE systems (and particularly in miniFACE systems). It is assumed that large fluctuations in pCO_2 have no effect on plant performance. There is now some evidence that this may not be the case as described in the following section.

Hileman *et al.* (1992) conducted a comparison of the pCO_2 distribution in a range of OTC designs, with and without a bevelled top edge, called a frustum. In three types of OTC (square, round without frustum and round with frustum) they found that variation over time was always greater than spatial variation within the chamber. Spatial variation in an OTC fitted with a frustum never exceeded 10% of the pCO_2 measured at the centre, but pCO_2 was frequently lower and more variable on the downwind side of the chamber. Drake *et al.* (1989) found pCO_2 to vary by 12% in an OTC design used in a tidal marsh community. This level of spatial variation is similar to that reported for FACE (Hileman *et al.*, 1992), although the effect of monitor specification, the use of a volumetric buffer (including sample line volume) and sample flow rate must be considered in relation to the short-term changes in concentration discussed in the next section.

B. Temporal Distribution of Gas Concentration

Summary information on the temporal control of pCO_2 in FACE systems is provided in Table 2 including data from small-scale FACE systems for comparison. The two fully operational BNL FACE facilities at Maricopa and Zurich have provided comprehensive data on the temporal distribution of pCO_2 (Nagy *et al.*, 1994, 1995). These systems proved to be able to control pCO_2 extremely well and to be highly reliable. In the MAC-FACE cotton experiments of 1990–1991, the 1-min average pCO_2 concentration was held within ± 10% of the set point > 90% of the time and ± 20% for > 99% of the experimental fumigation over two seasons. The system was not out of operation for more than 1% of the planned experimental period.

In the ETHZ-FACE experiments in 1993 and 1994, the facility operated for 99.3% ± 0.4% of the scheduled treatment hours (daylight hours in 1993 and daylight hours when leaf temperature was expected to be greater than 8°C in 1994), depending upon year and replicate. In both years, season-long average pCO_2 during scheduled daylight treatment hours were 59.8 ± 2 Pa at the control points. Instantaneous measurements of pCO_2 were within 20% of the target partial pressure of 60 Pa for 76–82% of the time when the facility was operating, depending upon replicate. One-minute mean pCO_2 values were within 10% of the target partial pressure for 89–94% of the time (Nagy *et al.*, 1995).

Table 2

Characteristics of temporal control of pCO_2 in FACE experiments

Vegetation	Plot Diameter (m)	Period of Measurement	Target pCO_2 (Pa)	Percentage of Observations with deviation from target of less than		Reference
				± 10%	± 20%	
Cotton	23/25	Apr–Sep	55	90	99	Nagy *et al.* (1994)
Ryegrass/clover	18	Jun–Nov and Feb–Nov	60	89–94	99	Nagy *et al.* (1995)
Dwarf wheat	1	Nov–Jul	60	> 90		Miglietta *et al.* (1996)
Limestone grassland species	0.1	4 months in summer	70	64		Spring *et al.* (1996)
Potato crop (at three different CO_2 levels)	8	Jun–Sep	66	66		Miglietta *et al.* (1997)
			56	74		
			46	83		
Loblolly pine	30	May–Oct	55	69	92	Hendrey *et al.* (1998)

All data for 1 min averaging time except for Spring *et al.* (1996) which used 15 min averages. The potential influence of the characteristics of sampling systems: CO_2 monitor specification, sample line length and diameter, and sample flow rate must be considered when evaluating the above data.

Control of pCO_2 in smaller FACE plots (e.g. 8 m and 1 m diameter; Tables 1 and 2) implies a smaller distance for gas mixing to occur and potentially larger fluctuations in pCO_2. A smaller plot size in relation to the scale of turbulence above vegetation also makes pCO_2 morc susceptible to turbulent eddies from above the canopy. These factors are likely to result in a greater variation of pCO_2 away from the target value in small systems (e.g. Miglietta *et al.*, 1997) compared to large-scale FACE systems (e.g. Nagy *et al.*, 1994, 1995) (Table 2). Characteristics of sampling systems, especially the use of long sample lines, low sample flow rates and a large volumetric buffer may all improve apparent performance characteristics and require consideration. Very small systems (e.g. 0.1 m diameter; Spring *et al.*, 1996) act by displacing the airflow around the canopy in a manner unlike large-scale FACE. The percentage time with deviations ± 10% from the target pCO_2 is significantly greater than with large-scale FACE and they are more susceptible to short-term fluctuations in pCO_2 and are discussed in the next section.

Temporal control within a FACE system in a forest canopy is more problematic than in a crop canopy because of the potential complications in airflow. The second Forest FACE prototype (FFP-2) (Hendrey *et al.*, 1994) achieved 1-min average pCO_2, measured at the plot central point, that was normally distributed ± 20% about the set point of 55 Pa for > 95% of operation during 38 12-hour days. The 5-min average concentrations were normally distributed ± 10% about the set point for > 99% of the time. Less than 2% of excursions outside ± 10% of the set point lasted longer than 60 s and < 3% outside ± 20% lasted as long as 60 s. The Forest system in a 10 m high canopy suggested a requirement of 1500 or 2000 metric tonnes CO_2 per 34 m diameter FACE plot for 12 or 24 h operation, respectively, or approximately 2300 kg per 12 h day for a seedling forest and wind speed of 1.5 m s^{-1}.

Although it is often assumed and stated (e.g. Rafarel *et al.*, 1995) that pCO_2 control in full-size FACE systems is not ideal, a detailed examination with other exposure methods can reveal a favourable comparison. Thus, the annual concentration observed in a "solardome" glasshouse facility, a closed-chamber system with constant airflow (Rafarel *et al.*, 1995) was 67.4 Pa for a set point of 69 Pa, which was 97.7% of the target value. This compares with 54.8 Pa for a target of 55 Pa (or 99.6%) over a growing season (April–September) in a FACE system (Nagy *et al.*, 1994). In three OTC designs, Hileman *et al.* (1992) reported fluctuations at all points sampled in excess of 10 Pa with a target pCO_2 elevation of 30 Pa.

1. Short-term Fluctuation in pCO_2

Variation of pCO_2 occurs in nature but it is likely to be more pronounced and frequent in experimental enrichment systems such as OTCs and FACE. An important issue that has received limited attention is the presence and any effect of very rapid short-term fluctuations in pCO_2 to which the plant is exposed but which may remain undetectable using conventional CO_2 monitors and sampling systems. Fluctuations are likely to be greatest close to the gas outlets. The need for a buffer zone between outlets and the usable area for biological sampling may exclude a significant area of small FACE plots and there appears to be a tendency to ignore adequate definition of these spatial gradients, concentration fluctuations and their possible impact.

It is difficult to conceive how release of gas from small jets within a short distance (< 1 m) can achieve adequate mixing before impacting the target vegetation. In reality, it is likely that pCO_2 fluctuates rapidly between the pCO_2 of the gas emitted from the sources and ambient pCO_2. Conventional sampling systems will not detect this because of the limitations caused by monitor response times, sample line volumes and sampling flow rates that will tend to buffer fluctuations. Thus, the measured concentration of 65 Pa achieved at the centre of a 10 cm diameter system (Spring *et al.,* 1996) is likely to fluctuate between the pCO_2 of 160–170 Pa (G. Spring, personal communication) emitted 5 cm away and ambient pCO_2. The significance of these fast fluctuations for plant physiology has yet to be fully evaluated but may be significant.

The effect of fluctuating pCO_2 on photosynthesis and transpiration have been examined by Cardon *et al.* (1995) who also reviewed earlier work and discussed their results in relation to several experimental methods. Cardon *et al.* (1994) showed that fluctuating pCO_2 could alter stomatal conductance far from the steady-state value established at the median pCO_2 and that the extent of the change was dependent upon the frequency of fluctuations and that there were species specific differences. These authors (Cardon *et al.,* 1995) further noted that fluctuating pCO_2 may influence transpiration, photosynthesis and water-use efficiency in both C_3 and C_4 species, and that the fluctuations in an elevated pCO_2 experiment may constitute a treatment separate from any increase in mean pCO_2. The fluctuations may confound comparisons of water-use efficiency and the competitive ability of plants grown at different mean pCO_2 concentration. They reported the occurrence of fluctuations of 150 Pa in 5-s mean pCO_2 in an OTC receiving 35 Pa above ambient pCO_2 but were unable to determine a particular frequency of fluctuation.

Hendrey *et al.* (1995, 1997) undertook experiments to determine whether photosynthesis in wheat (*Triticum aestivum* cv. Hereward) could sense fluctuations in pCO_2. They found that symmetrical oscillations of 22.5 Pa about a set point of 57.5 or 65 Pa produced no effect on instantaneous chlorophyll fluorescence compared to a constant concentration when half-cycle changes in pCO_2 were less than 2 s (Hendrey *et al.,* 1997). However, at fluctuations of longer than 2 s, the fluorescence signal also cycled in response to the changes in pCO_2. After suppressing photosynthesis with a high concentration of oxygen, they also found that linear electron transport was decreased by half-cycles of 60 s or longer. They concluded that such oscillations of 1 min or longer decrease photosynthetic uptake of CO_2 in wheat and that evaluation of results from CO_2-enrichment experiments would require careful assessment of short-term variation in pCO_2.

There still remain important questions to be resolved about the magnitude and variability of such responses between species. At present, many FACE experiments do not provide adequate quantification of short-term variation in pCO_2.

C. FACE Blower Effect

The nature of unfumigated control plots in the MAC-FACE experiments have evolved during the course of the research programme. Initially, control plots had no surrounding plenum and pipes, but in the FACE-cotton experiment of 1990 onwards, a "false" plenum was introduced in the MAC experiments surrounding the control plots, although a blower was not used (Mauney *et al.,* 1994). However, the results from the FACE wheat experiments produced an unexpected observation. Canopy temperature rose at night as well as during the day and this could not be explained by a decrease in latent heat transfer (Kimball *et al.,* 1995). Although the air blown into FACE rings contributes a negligible amount to the normal air flow, under still conditions it can act to draw in warmer air from higher in an inversion and blow this over the crop. Climatic inversions are common at night at the MAC site and blower air circulation was suggested as the cause for increased canopy temperature at night. In the 1996 and 1997 wheat experiments, this artefactual source of warming was tested by adding blowers to the "control" plots, and adding further plots which included the FACE ring structure, but not the blowers (termed "ambient" plots). In 1996, it was found that FACE and control plots, both now with blowers, continued to show significant differences in temperature during the day which could be attributed to inhibition of transpiration by elevated pCO_2, but there was no CO_2 effect at night. The FACE and control plots were, however, significantly warmer than the ambient plots at night, confirming that the earlier warming at night was an effect of the FACE system on microclimate under inversion climatic conditions and was not caused by elevated pCO_2 (Garcia *et al.,* 1998). This presents a dilemma of whether to fumigate at night or not. Fumigation at night will ensure that any effect of elevated pCO_2 on respiration (Drake *et al.,* 1997) is included, but with the risk that temperatures will be elevated under inversion conditions when pCO_2 would naturally rise well above that of the bulk atmosphere. If fumigation at night is included then this suggests that additional plots without blowers, as in the Maricopa experiment, will be necessary to identify the blower effect.

This effect of FACE blower air flow on canopy temperature (and potentially also on pCO_2 within the canopy under inversion conditions) constitutes a microclimatic modification of FACE methodology or a "chamber effect" which must be considered alongside the changes to microclimate imposed in OTC studies.

D. FACE Island Effect

FACE plots are isolated "islands" of elevated pCO_2 within an experimental field and this effect may have physical and biological consequences (McLeod, 1993). The humidity and vapour pressure deficit experienced by vegetation is influenced by the transpiration of the surrounding areas. At ambient pCO_2, the vegetation surrounding a FACE plot may be transpiring at a higher rate than within the elevated pCO_2 plot. This potential of the "island" effect on water vapour fluxes has not been assessed in the literature.

The "island" effect may also influence some biological observations. FACE plots may appear as foci for attacks by mobile herbivorous insects or pests, which may become unusually attracted and concentrated within plots, achieving immigration rates which would not occur if the surrounding area was also at elevated pCO_2. Consequently, the observed effects may not be representative of the effect of globally elevated pCO_2. Conversely, conditions leading to a more rapid population growth of some organisms (e.g. insects or fungi) within plots, arising from treatment effects, may be less than would occur with globally elevated pCO_2 because of loss of propagules from the plot (e.g. fungal spores) which are not recruited from the surrounding areas. These potential limitations of FACE appear less severe than the restrictions on pest and pathogen movement, and the changes in microclimate imposed by the structure of OTCs. However, the importance of the "island" effect requires evaluation for particular biological observations in FACE experiments, such as the population dynamics of insect herbivores and their parasites that may be scale-dependent.

E. Costs of FACE Facilities

FACE systems use large amounts of CO_2 gas and this can be a major contribution to their running costs. Consequently, there has been considerable interest in alternative sources of CO_2 (see e.g. Allen, 1992) from geological sources (e.g. Miglietta *et al.*, 1993) and from industrial processes. Geological sources of CO_2 are restricted to certain areas of the world (Koch, 1993). These and industrial sources must be carefully evaluated for contaminant gases which may confound responses to elevated pCO_2.

Carbon dioxide use in a FACE experiment is influenced by canopy structure, wind velocity and atmospheric stability, the latter itself influenced by incident solar radiation. Very windy sites would use large quantities of CO_2. Typical CO_2 usage on 23 m diameter FACE arrays at Maricopa between 20 May and 18 September 1989 was 1400 kg CO_2 per day per array between 0600 and 1800 MST in order to achieve a target concentration of 55 Pa (Nagy *et al.*, 1992). They reported that CO_2 use was not strictly proportional to wind speed because vertical dispersion plays an important role in the CO_2 requirement. Increased vertical turbulence and vertical "loss" of CO_2 from the fumigated plot increases the CO_2 usage required to achieve a given target elevation. Thus, diurnal changes in CO_2 use per unit wind speed at Maricopa reveal highest values at midday when the intense Arizona sun creates atmospheric turbulence. Nagy *et al.* (1992) also report that rapid changes in wind direction and speed, as indicators of turbulence, also correlate with changes in CO_2 usage. They provided a formula for calculating CO_2 use for a 23 m BNL FACE array:

$$F = 0.3 + 1.1\,u + 0.8 \sin(h) \qquad (1)$$

where F is the rate of CO_2 use in kg min^{-1}, u is the wind velocity in m s^{-1} and h is the solar elevation above the horizon.

The average CO_2 use in the BNL Forest FACE system with a 10 m tall canopy (Table 1) was 260 kg h^{-1} in the 26 m diameter plot or 50 g h^{-1} m^{-3} of forest volume (Hendrey *et al.*, 1998). Apsley (1991) reported theoretical calculations for a FACE facility for young trees in SE England where the mean annual wind velocity at 2 m height was 2.2 m s^{-1}. These suggested an annual CO_2 usage rate of approximately 4.5 tonnes m^{-2} (varying slightly with plot size) in order to achieve 54 Pa with the atmospheric stability conditions typical of the site.

The use of a dual-ring FACE system in which an outer array of gas jets was directed outwards in an attempt to reduce mean wind velocity and thereby reduce gas use has also been investigated (Reece *et al.*, 1995). However, such a barrier of outward facing jets has the potential to entrain CO_2-enriched air in the jets and to form a barrier to the bulk airflow that increases turbulence over the plot. Both effects would increase CO_2 use. Modelling of this approach using the computational fluid dynamics methods of Walklate *et al.* (1996) suggests that CO_2 use may be slightly increased compared to conventional FACE designs that use only inward-facing jets (P. Walklate, personal communication). The dual-ring approach has not been adopted in any current FACE experiments.

It should be noted that calculations of CO_2 usage rates for the purpose of designing FACE facilities should take account of the frequency of maximum CO_2 demand at high wind velocities, the storage tank capacity, the rate of replenishment of CO_2 supplies and the maximum achievable rate of liquid CO_2 evaporation. Site characteristics such as the height of the vegetation, topography and location (which influences atmospheric stabilities) must also be considered. All these factors influence the design and costs of a FACE installation. Liquid CO_2 stored in tanks has been used in most FACE facilities that usually require a small refrigeration plant to maintain the low storage temperature. This has an electricity requirement with a recurrent cost. A further requirement is a means to provide the latent heat for evaporation to the gaseous state. In some fumigation facilities (e.g. Rafarel *et al.,* 1995) electrical evaporators are used with a further recurrent electrical cost. In large FACE systems such electrical costs may be a large component of recurrent running costs to be considered in addition to the cost of CO_2 gas. However, the FACE system at Maricopa utilises the water from an adjacent irrigation ditch as a source of heat for evaporation, thus avoiding electricity costs. In the FACE systems at Eschikon and Duke Forest, an ambient air temperature evaporator is used as the source of heat.

The costs for building and operating BNL FACE systems, including some comparisons with closed and open-top chamber systems have been provided by Kimball *et al.* (1992) and Hendrey *et al.* (1993). In addition, Schulze and Tenhunen (1993) provided costs for a proposed FACE system in a mixed forest in Germany. Scientists from the BNL, USA, were involved in establishing the Swiss FACE system, for which costs estimates are provided by Blum (1993).

In a comparison of costs between FACE and other CO_2-enrichment methods, Kimball *et al.* (1992) compared the BNL FACE experiment at Maricopa with an OTC study and sunlit controlled environment chambers (soil, plant, atmosphere research chambers: SPAR) of Jones *et al.* (1984) (Table 3). Each method assumes 16 plots of which eight are enriched to 30 Pa above ambient for 6 months. Enriched areas are 16, 72 and 874 m^2, respectively. (See Kimball *et al.* (1992) for full details of assumptions and units costs). Kimball

Table 3
Cost comparisons ($000) between SPAR, OTC and FACE methods of CO_2 enrichment (from Kimball, 1992)

	SPAR	OTC	FACE
Initial setup costs	350	200	300
Annual running costs	129	128	438
(of which CO_2 costs)	0.12	25	288
CO_2 cost per m^2 enriched area ($ m^{-2})	7.5	347	329
CO_2 and other costs per m^2 enriched area	9.3	1.8	0.5

Each method assumes 16 plots of which eight are enriched to 300 ppm above ambient for 6 months. Enriched areas are 16, 72 and 874 m^2, respectively. See Kimball (1992) for full details of assumptions and unit costs.

et al. (1992) noted that the major costs of CO_2-enrichment experiments are the scientific staff and instruments required to support biological studies and he estimated the costs of producing treated plants using enrichment facilities as only 26%, 18% and 28% of total project costs for SPAR, OTC and FACE methods, respectively. Consequently, it is important to consider carefully the common perception that FACE methods are the most expensive enrichment technique against the scientific and financial advantages of scale. An analysis of costs per unit area with elevated pCO_2 reveals a much lower cost of FACE compared to other enrichment methods (Table 3; Kimball *et al.*, 1992).

V. THE POTENTIAL AND REALISED CONTRIBUTION OF FACE TO BIOLOGY

Given the high absolute costs of a FACE facility, what benefits can the technique provide in understanding the biological effects of rising pCO_2 that are not already available from much cheaper, more easily operated and more widely available enclosure systems such as OTCs? This section reviews the theoretical advantages of FACE with respect to biological questions, highlights the new findings emerging from the full-size FACE experiments and points to future issues that FACE experiments should resolve.

In section II we outlined the limitations imposed by OTCs for predicting the effects of elevated pCO_2 on vegetation. The result of these is that chamber effects are often as large or larger than those imposed by the treatment itself (Fuhrer, 1994; Janous *et al.*, 1996). All major OTC experiments provide chambered and open-air controls, the assumption being that, whilst environmental conditions inside the chambered control differ from those of the open-air control, the difference between the chambered control and chambered treatment reflects the difference that would occur between the open-air control and the hypothetical open-air treatment. Is this assumption valid? This section examines the potential and actual problems of predicting effects of elevated pCO_2 on different vegetation variables from OTC studies and the extent to which FACE has or may overcome these.

Although a large number of FACE experiments are now in operation, published data are largely limited to experiments on four crop/vegetation types at three locations:

(1) FACE with cotton at Maricopa, USA (section III.B.1);
(2) FACE with wheat at Maricopa, USA (section III.B.2);
(3) FACE with a variety of species, primarily ryegrass/clover at Eschikon, Switzerland (section III.B.3);
(4) FACE with a loblolly pine overstorey at Duke Forest, USA (section III.B.4).

These form the focus of the following discussion.

A. Production, Biomass and Efficiency of Light Use

Biomass represents a major pool of C in the biosphere, similar in magnitude to that constituted by CO_2 in the atmosphere. A major focus of studies of natural ecosystems is therefore the potential for change in the biomass pool with rising pCO_2. Similarly, for crops the question of whether yields will increase, by how much and under what conditions, is critical to predicting food security into the twenty-first century. Addressing both issues requires the most realistic estimates of plant responses to elevated pCO_2 that are possible.

1. Limitations of OTCs and the Theoretical Advantages of FACE

The database of elevated pCO_2 effects is based largely on measurements made in OTCs (reviewed by Kimball, 1983; Drake *et al.,* 1997) and therefore in an aerial microclimate that is duller, warmer and drier than the norm (see section III). Since controls replicate these conditions, does this matter? Temperature and light modify the acclimation of plants to elevated CO_2 (Ghannoum *et al.,* 1997; Kellomaki and Wang, 1997; Read *et al.,* 1997; Stirling *et al.,* 1997; Vu *et al.,* 1997; Ziska and Bunce, 1997; Curtis and Wang, 1998). Although this chamber effect is likely to be too small to cause a change in the direction of a response, it will alter its magnitude. Additionally, small isolated plots are well known, in agronomic trails and ecological experiments, to overestimate biomass, production and yields (Roberts *et al.,* 1993). Increased radiation interception at the edges of small plots, in particular, can exaggerate the effect of a treatment. Open-top experiments and some closed-chamber experiments have suggested very large increases in the yields of C_3 crops under elevated CO_2 (Kimball, 1983). The maximum practical size of OTCs (typically < 2 m diameter) limits each plot to a ground surface area of about 3 m^2. In a 2 m diameter chamber, more than 50% of the vegetation will be less than 30 cm from the chamber wall and 75% within 50 cm of the wall. The plot sizes used in agricultural trials usually include a border or buffer area of twice the vegetation height (Roberts *et al.,* 1993). Therefore, even a 50 cm high semi-dwarf wheat crop would require a buffer zone of 1 m and no area within the OTC would be free of edge effects. Consequently, knowledge of crop responses to elevated pCO_2 is currently derived from experiments that would be considered unacceptable in agronomic trials.

For forests, which are estimated to comprise 80–90% of the global pool of total C in living plants (Melillo *et al.,* 1990), the situation is more deficient. OTCs can only accommodate one or two moderately sized trees, and therefore edge-effects are likely to be extreme and natural canopy closure is prevented. As noted by Lee and Jarvis (1996) "trees . . . do not fit into simple experimental enclosures. Furthermore, trees and forests are very well coupled to the atmosphere, and this coupling is often greatly reduced when trees are enclosed in chambers, introducing an additional artefact". Experiments with trees, with a few exceptions, have been limited to seedlings and juvenile trees (Lee and Jarvis, 1996). Given that juvenile trees can respond very differently to light when compared to mature individuals, it is similarly likely that responses to pCO_2 will also be misleading about the effect on mature trees (Lee and Jarvis, 1996). Studies with seedlings are of value in establishing impacts at the establishment phase. For tall trees, in individual chambers, root systems may often spread beyond the area covered by the chamber, unless restricted by an artificial physical barrier. Such a barrier will solve the problem of roots exploiting soil outside the chamber, but may artificially induce a feed-back inhibition on photosynthesis and production (Arp, 1991). Branch bags have allowed investigation of the response of individual branches of mature forest trees. However, loss of photosynthetic capacity in acclimation under elevated pCO_2 is assumed to result from increase in the ratio of the capacity of the plant to produce photoassimilate (source strength) to its ability to utilise the carbohydrate formed (sink strength) (Drake *et al.,* 1997). This increase in source : sink ratio is reflected in an accumulation of photoassimilate in the source leaves which results in repression of so-called "feast" genes. These include the *rbcS* gene family which code for the small sub-unit of the primary photosynthetic carboxylase, Rubisco, in addition to other genes coding for proteins of the photosynthetic apparatus (Koch, 1996; Drake *et al.,* 1997). Whilst it may be argued that tree branches enclosed in branch bags are autonomous at maturity, export will depend on the non-structural carbohydrate reserves of the rest of the plant. This is likely to be very different when the rest of the plant is also growing in elevated pCO_2. Similarly, responses have often been linked to prolonged soil moisture conservation under elevated pCO_2. The supply of water to the elevated pCO_2 branch may therefore be very different to that which would occur if the whole plant had been grown in elevated pCO_2.

Theoretically, the edge effects of OTCs, in terms of light interception, may be minimised if the chamber is surrounded by identical vegetation. However, access to the chamber edge, trampling in the construction of the chamber and the effects of wind lodging vegetation around the chamber mean that this can rarely be achieved. The chamber walls themselves constitute a further edge effect by restricting the canopy and by alteration of light quantity, quality and its angular distribution. In long-term experiments, any increase in growth of the vegetation inside the chamber relative to that outside will further increase the effective edge intercepting radiation and could lead to a positive feed-forward effect on production, particularly for trees. For arable crops, large FACE systems, such as those used for cotton and wheat at Maricopa overcome most of the edge effects. At maturity, for the semi-dwarf cultivar of wheat grown in 1993, 83% of the crop would be outside the area considered to be affected by edge-effects, compared to 0% for a 2 m OTC. In trees, however, even the 30 m diameter ring of the FACTS-I with 12 m tall trees (Hendrey *et al.,* 1998) may not fully avoid edge effects. This though is mitigated by the fact that erection of this FACE facility appears to have occurred without any significant interruption of the forest canopy (Figure 1).

Photosynthesis and therefore efficiency of light energy conversion at the leaf level is almost invariably observed to increase in elevated pCO_2 (reviewed by Gunderson and Wullschleger, 1994; Sage, 1994; Drake *et al.,* 1997). If this is translated into increased production at the stand level, then the efficiency of conversion of intercepted light into biomass (ϵ_c) should also increase. If increased production results in increased leaf area or more rapid canopy development then the efficiency of light-interception (ϵ_i) will increase. Understanding of how these variables will increase under elevated pCO_2 is of particular importance, since they form a basis of landscape, regional and global models of vegetation response to rising pCO_2 (Amthor, 1994). Both variables require measurement of radiation interception within the stand. Testing this question has been difficult to achieve until FACE, since light absorption in field OTCs is increased by penetration of light through the disturbed edges and by depression of the proportion of direct light at low sun-angles by the chamber walls.

2. *Findings within FACE experiments*

The FACE experiments with cotton and wheat crops provided the first clear evidence that elevated CO_2 significantly increases ϵ_c. The scale of these FACE experiments allowed regular harvests of sub-plots within each of the four replicate blocks at regular intervals (< 2 weeks) throughout the duration of the crop. This was coupled with regular measurements of light interception by the canopy, allowing a record of both ϵ_i and ϵ_c throughout the growing season. In cotton, elevated pCO_2 resulted in significant increase in ϵ_i during the early and middle stages of crop development, but this increase was lost in the final stage of crop growth. However, ϵ_c was increased throughout from 0.0265 (= 2.65%) to 0.0335 (calculated from Pinter *et al.,* 1994). The value of ϵ_c for the control crop was close to the 0.028, suggested by Monteith (1978) as a maximum for healthy C_3 crops in the current atmosphere. This result provides clear evidence that elevated pCO_2 increases the efficiency of the crop above an apparent ceiling for C_3 photosynthesis in the current atmosphere. The increase in ϵ_c was less than the increase in photosynthesis at the leaf (*c.* 30%) and canopy (19–41%) level (Hileman *et al.,* 1994; Idso *et al.,* 1994), but may be accounted for by a larger increase in root production (Prior *et al.,* 1994b). The increase in ϵ_c corresponded to a cotton boll yield increase of 43% (Mauney *et al.,* 1994).

In wheat the yield of a fully irrigated crop was increased significantly by 8–12% (Kimball *et al.,* 1995). The difference between wheat and cotton is consistent with the much greater stimulation of photosynthesis by elevated pCO_2 expected at the warm summer temperatures under which cotton grows versus the cooler spring conditions under which wheat grew at this site (Long and Drake, 1991; Drake *et al.,* 1997). In the experiments at Maricopa with wheat there were significant interactions between elevated pCO_2 and soil moisture treatments. In wheat, the increase in yield with elevated pCO_2 was 8–12% with full irrigation but 21–25% in the drought treatment (Kimball *et al.,* 1995; Pinter *et al.,* 1996).

In the FACE experiment at Eschikon, the yield of clover in monocultures increased in the first 2 years by 20%, declining to 11% in the third year. The response in all years was independent of defoliation frequency and nitrogen fertilisation. In ryegrass monocultures, yield increased by only 7% on average over 3 years at elevated pCO_2, and varied according to defoliation frequency and nitrogen fertilisation. There was no significant increase in yield in elevated pCO_2 in the low nitrogen treatments, even though photosynthesis was increased significantly in both treatments by 35–40% (Rogers *et al.,* 1998). Although yield of ryegrass was not increased by elevated pCO_2 in the low nitrogen treatment, the amount of dead vegetation in the plots did increase significantly (Blum *et al.,* 1997), demonstrating increased turnover and net primary production. This effect has been suggested as a possible explanation for the mismatch between increased CO_2 uptake in elevated pCO_2 relative to often smaller increases in biomass (Hungate *et al,* 1997). In the bi-species mixture, the increased response to elevated pCO_2 by clover was reflected in an increase in the proportion of this crop within the mixture (Hebeisen *et al.,* 1997a,b).

Both the Maricopa and Eschikon FACE experiments have revealed significant changes in the quality of the harvested product. Clover herbage showed a significant 4% decline in protein content, but no change in the proportion of digestible fibre, energy for lactation or cyanogenic glucosides (Frehner *et al.,* 1997). Wheat stems grown in elevated pCO_2 showed an increased cellulose and decreased lignin content, with lower protein contents (Akin *et al.,* 1995a).

The FACE experiment with loblolly pine has been in operation for too short a period to assess direct effects on production.

3. How Has FACE Influenced Understanding?

The completion of crop trials at scales acceptable in agronomic trials has increased confidence in the predictions of yield increases under elevated pCO_2. The expectation prior to these experiments was that elevated pCO_2 would be less effective at promoting crop yields under stress conditions. This expectation has been a key factor in predicting security of global food supplies under climate change (Rosenzweig and Parry, 1994). The contrary findings of these FACE experiments now demand a re-evaluation of this issue.

The increase in cotton yield was similar to that predicted from OTC experiments (Pinter *et al.,* 1996). However, the 8–12% increase in wheat yield at 55 Pa at Maricopa compares to an average 40% increase in wheat yield across an OTC network study using 70 Pa (Hertstein *et al.,* 1996). This implies a minimum increase of about 20% at 55 Pa and about double the increase actually observed in FACE. These results were obtained in different years and at different locations, so it may be difficult to account for the disparity. However, it does raise the possibility of overestimation in the OTC studies by, for example, edge effects. OTC studies have produced very different results with the same crop at different sites (Hertstein *et al.,* 1996). Since, as yet, no single crop has been examined at more than one location with FACE it cannot be said with certainty whether the productivity results obtained with FACE are part of this continuum of variation or a different answer. The observation in Eschikon that the response of yield to elevated $p\mathrm{CO_2}$ may decline over 3 years highlights a further advantage of FACE. Such a decline using OTCs could result from a cumulative chamber effect interacting with the $p\mathrm{CO_2}$ treatment. For example, the long-term result on a perennial crop of growth at lowered light levels or in decreased rainfall imposed by the chamber.

It can be said with certainty that FACE is at an appropriate scale for field trials of crops such as wheat and avoids the major artefacts imposed by other exposure methods, at least for herbaceous crops. The technology and its durability have been proven at the Maricopa and Eschikon sites (Lewin *et al.,* 1994; Nagy *et al.,* 1994; Hebeisen *et al.,* 1997b). In order to give any certainty to predictions of how a single food crop will respond to elevated $p\mathrm{CO_2}$, a network of a minimum of 4–5 FACE sites may be required to accommodate the variation in climate and soil conditions across the geographical range of the crop.

B. Water and Energy Balance

With the exception of some major tree species, such as *Fagus sylvatica* (Heath and Kersteins, 1996) and *Pinus taeda* (Ellsworth, 1998), most terrestrial plants respond to elevated $p\mathrm{CO_2}$ with a reduction in stomatal conductance. This is maintained in the long term and in some cases may be reinforced by a decreased stomatal index and photosynthetic acclimation (Sellers *et al.,* 1996; Drake *et al.,* 1997). Across studies of a wide range of species and conditions, leaf area index (LAI) was found, on average, to vary insignificantly with growth at elevated $p\mathrm{CO_2}$ (Drake *et al.,* 1997). If stomatal conductance decreases, whilst surface area and canopy structure remain unchanged, then transpiration and the ratio of latent to sensible heat transfer from canopies should decline and result in surface warming (Sellers *et al.,* 1996, 1997). The extent of this response increases with canopy–atmosphere coupling. This may be further emphasised by the asymmetrical response, whereby the stomata of the upper surface of amphistomatous leaves can show a stronger closing response at elevated pCO_2. Under conditions of natural convection this would result in a higher ratio of sensible to latent heat loss than with symmetrical closure (Pearson *et al.,* 1995).

1. Limitations of OTCs and the Theoretical Advantages of FACE

In addition to the problem of scale, OTCs present problems in attempting to test the magnitude of any decrease in stand transpiration and increase in temperature. By sheltering the vegetation from the wind and by continuously forcing air across or through the canopy, and upward out of the chamber, an OTC alters the coupling of vegetation with the bulk atmosphere in a complex way. In the open air, coupling of canopy elements will normally improve upwards towards the top of the canopy. In OTCs, air is often blown into the chamber at or below the top of the canopy. This air movement will commonly disturb the base of the canopy, so potentially improving the coupling of the lower canopy foliage and soil surface with the atmosphere. Thus, chamber effects on water loss will be complex. The effects of stomatal closure on water loss will be exaggerated compared to the situation in the open air if coupling is improved and diminished where coupling is degraded. The "chamber effects" will therefore be complex, variable within a canopy and dependent on the weather. Dense low vegetation, such as short grassland, is normally poorly coupled to the atmosphere. Thus, coupling might be improved for this vegetation by the air flow in OTCs. An analysis of gas exchange in an OTC by Unsworth *et al.* (1984) revealed that leaf boundary layer resistances of soybean were small inside an OTC compared with values in crop canopies where wind speeds seldom exceed 1 m s^{-1}.

Pleijel *et al.* (1994), using a standard OTC design to study an oat crop, noted that O_3 deposition in the canopy inside the OTC over a period of time was significantly greater than outside, suggesting that coupling with the canopy was improved in the OTC. Even where coupling is unchanged relative to outside, increased air temperature will raise the vapour pressure deficit. This effect could be substantial. Consider air above a canopy and forced into an OTC with a dew point of 10°C and an air temperature of 14°C in the sun. If a 3°C rise in air and canopy temperature occurs in the chamber, then the vapour pressure deficit would rise by over 110%, and thus the potential evaporation would more than double simply because of the presence of the chamber. If this occurred, the rate of water loss in control OTCs would be more than double that of the same vegetation in the open air. Any effect of elevated pCO_2 on canopy transpiration would be exaggerated in proportion, resulting in much larger differences than could occur in the open air. For trees, which are normally well coupled to the atmosphere, OTCs are likely to degrade coupling (Lee and Jarvis, 1996), except under very still atmospheric conditions. Here the effect of elevated pCO_2 on water loss may be underestimated in OTCs. Determining whether elevated pCO_2 decreases water loss from the system will be further complicated in OTCs because of lateral migration of moisture into the adjacent drier soil outside of the chamber. It will also be complicated by the impact of the chamber wall on the distribution of precipitation and by any experimental protocol which provides artificial watering to maintain soil saturation in both control and treatment chambers. For these reasons, transpiration and energy balance were major areas of investigation in the cotton and wheat crops grown in the FACE experiments at Maricopa (Kimball *et al.*, 1994, 1995; Pinter *et al.*, 1996).

2. *Findings within FACE Experiments*

Measurements of leaf and canopy evaporation, stem flow, soil moisture, water balance, energy exchange and canopy temperature, in crops of cotton and wheat were made throughout the FACE experiments at Maricopa. Stomatal conductance of the fully irrigated cotton crop was only decreased by elevated pCO_2 at the end of the growing season, but decreased in the droughted crop throughout (Bhattacharya *et al.*, 1994). Elevated pCO_2 only caused increased shoot water potential at the end of the growing season (Bhattacharya *et al.*, 1994). In wheat, elevated pCO_2 decreased stomatal conductance throughout the growing season (Garcia *et al.*, 1998), although stomatal density was unaffected (Estiarte *et al.*, 1994). These decreases at the leaf level corresponded to significant decreases in stem water flow and canopy transpiration in both cotton (Dugas *et al.*, 1994; Hileman *et al.*, 1994; Kimball *et al.*, 1994) and wheat (Kimball *et al.*, 1995; Senock *et al.*, 1996). Although calculation of the energy balance suggested a decrease in latent heat transfer for the cotton crops, the cumulative decrease in latent heat loss estimated by difference from the energy balance equation was not statistically significant (Kimball *et al.*, 1994). Use of more sensitive radiometers in the wheat experiments showed statistically significant 8% and 11% decreases in cumulative latent heat transfer and therefore ET_c over two consecutive growing seasons (Pinter *et al.*, 1996). A 5% decrease in ET_c was indicated from soil moisture depletion over the same years, but this was not statistically significant. In the droughted plots there was no evidence of a decrease in ET_c (Hunsaker *et al.*, 1996; Pinter *et al.*, 1996). The difficulty in detecting an effect of elevated pCO_2 in this FACE experiment compares sharply and disturbingly with the large positive effects on soil moisture noted in OTC studies in similarly dry environments (e.g.

Field *et al.*, 1997; Owensby *et al.*, 1997). Field *et al.* (1997) found a 50% decrease in evapotranspiration leading to equivalent improvements of soil moisture with a doubling of pCO_2 (Fredeen *et al.*, 1997). Interestingly, both studies concern grasslands where natural coupling between canopy and the atmosphere is poor relative to forest. This again raises the questions of whether the decreased community water loss in elevated pCO_2 when comparing control and treatment OTCs could at least in part be an artefact resulting from altered plant–atmosphere coupling in both control and treatment OTCs.

The decrease in transpiration of the wheat experiment at Maricopa also corresponded to a warmer canopy observed independently from hand-held, pylon and aircraft-mounted sensors. Daytime canopy surface temperature increases observed from aircraft remote sensing due to elevated pCO_2 were 1.2°C in the fully irrigated sub-plots and 1.7°C in the droughted sub-plots. Over the major growth period of the crop, the surface temperature was increased by an average of 0.6°C (Kimball *et al.*, 1995).

As discussed previously (section IV.C), part of this temperature increase is almost certainly an artefact of FACE. That is, under the still-air conditions that can occur at night and which allow temperature inversion, the energy introduced by the blowers is sufficient to break down the inversion, bringing warmer overlying air to the surface. Additional warming would increase evaporation, and could have created an artificially small difference in evapotranspiration between controls and FACE. Since transpiration will occur almost entirely during the day, when still conditions would be very rare, it is unlikely that this artefact could have any substantial effect.

In summary, a decrease in evapotranspiration for wheat of 5–10% was calculated independently from mass balance, canopy gas exchange and energy balance to close agreement, although not all of these differences were significant. In addition, the canopy gas exchange only covered short periods of the total growing season. The extent to which the "blower effect" may have decreased the difference is being partly revealed from the experiments in 1996 and 1997 when blowers were added to the control rings. Here a difference between the canopy temperatures of the elevated and control rings apparently remains during the day, but not at night (B.A. Kimball, personal communication). However, this addition of blowers cannot resolve the question of a blower effect fully. First, the solution has been to add the same artefact to control and elevate pCO_2 treatment. Secondly, the treatment differences are confounded with years of measurement. To address this, two further controls without blowers were added to the experimental design on the north side of the field in 1996 (S.P. Long, personal observation). In addition to the possibility of confounding subtle site differences with these controls because of their location, the statistical power provided by such a small replicate number could only detect very large differences.

Highly significant decreases of stomatal conductance averaging *c.* 40% across the growing season have been observed in both ryegrass and clover in the FACE experiment at Eschikon (Bryant, 1994; Lewis, 1994; Nijs *et al.*, 1997). In common with other evergreen conifers, the loblolly pine of the Duke Forest FACE experiment appears to lack any direct response to elevation of pCO_2. As a result, neither needle transpiration nor plant water relations appear to change under elevated pCO_2 (Ellsworth *et al.*, 1995; Ellsworth, 1998). Although soil water balance is being measured in both experiments, the results have yet to be reported.

3. How Has FACE Influenced Understanding?

Sellers *et al.* (1996, 1997) have carefully reasoned that the anti-transpirational effect of elevated pCO_2 on vegetation will cause continental warming of a similar magnitude to that predicted for increased heat retention by the global atmosphere, owing to increased greenhouse gas concentrations. This may denote the most important feedback of vegetation on the global climate system under atmospheric change. By scaling the decrease in stomatal conductance induced by elevated pCO_2 to the regional level and coupling the changed energy balance to a layered global circulation model, Sellers *et al.* (1996) showed that surface temperatures would rise by 1–2°C. What experimental evidence is there for such an increase? The aerial false colour thermal images of the 25 m FACE wheat plots at Maricopa during the daytime must represent the best and most convincing data available so far. These show a temperature increase of 1.2–1.7°C (Kimball *et al.,* 1995). The significant increase in sensible to latent heat exchange and the persistent decrease in stomatal conductance with no significant change in canopy size (Pinter *et al.,* 1996; Garcia *et al.,* 1998) are all consistent with the model of Sellers *et al.* (1996). Many OTC results have shown similar or larger decreases in transpiration, but their scale and the chamber walls would prevent the application of the energy balance approach used in the FACE experiment at Maricopa. Altered coupling, altered soil moisture and chamber atmosphere humidity also added great uncertainty to the significance of any observed temperature increase in OTCs. The plot scale (380 m^2) and the large range of measurements, in the absence of chamber effects, in the Maricopa FACE experiment remove the uncertainties of this major effect of pCO_2 on surface temperatures. However, the results from the Duke Forest experiment also suggest that a significant part of the terrestrial biosphere, i.e. the boreal forests, may not have the potential to produce such warming. Thermal images, such as those produced for the Maricopa experiments (Kimball *et al.,* 1995), will consequently be of great value from other FACE sites in evaluating the significance of the predictions of Sellers *et al.* (1996).

C. Effects on Photosynthesis, Metabolism and Molecular Biology

Decreased photosynthetic potential and its significance to net primary production has been a major area of debate in projecting the long-term effects of elevated CO_2 on ecosystems and crop production (Drake *et al.*, 1997). Whilst the view that decreased photosynthetic potential would offset the initial gain owing to decreased photorespiration has been widely publicised (Oechel *et al.*, 1994), others have argued that this is rare, especially in the absence of artefactual restrictions on growth (Arp, 1991; Sage, 1994). In addition, others have noted that acclimation can represent a reallocation of nitrogen away from the major leaf enzyme Rubisco, which has increased efficiency under elevated CO_2 (Webber *et al.*, 1994; Woodrow, 1994a,b). Such re-optimisation of limiting resources would increase rather than decrease production under elevated pCO_2 (Drake *et al.*, 1997). A major problem in assessing the significance of acclimation has been the spatial and temporal variability in observed acclimation, even within an intensively studied species such as wheat. Differences may simply result because studies have been undertaken at different stages of development (Nie *et al.*, 1995b). Differences may also result from different methods of elevating pCO_2 when these are associated with other changes such as restriction of rooting volume, depression of ambient light or elevated temperature (Sage, 1994; Drake *et al.*, 1997).

1. Limitations of OTCs and the Theoretical Advantages of FACE

Acclimation of photosynthesis is a dynamic process depending on the time of year, the stage of canopy and leaf development, and is explained at the level of gene expression (Stitt, 1991; Sheen, 1994; Nie *et al.*, 1995a; Koch, 1996; Drake *et al.*, 1997). Understanding whether acclimation is a transient phenomenon and understanding the mechanism of acclimation, requires repeated analyses of photosynthesis of leaves at different stages of development. It also requires harvesting of significant quantities of leaf for protein, metabolite and nucleic acid analysis. If we assume that no more than 5% of a canopy should be removed by destructive analyses within a year, this would provide just 400 cm^2 per unit canopy layer of LAI in a 1 m diameter OTC, compared to 250 000 cm^2 per unit canopy layer in the 25 m diameter FACE plots of the Maricopa experiments (sections III.B.1–2). An area of 400 cm^2 would limit several biochemical and molecular biological analyses and may not allow repeated sampling. Repeated sampling would be essential, for example, to understand the seasonal progression of photosynthetic acclimation or the yearly progression in long-lived foliage. Similarly, changes in secondary metabolites will be dynamic and can only be meaningfully studied by analysis of the time sequence through frequent sampling. For example, Nie and Long (1992) showed large differences in the amounts of a specific glycoprotein induced by growth in elevated pCO_2 in sour orange trees during the autumn and winter that were not apparent in spring and summer.

Measurements of leaf photosynthesis *in situ* have been confined, almost exclusively, to the upper leaves of plant canopies in elevated pCO_2 research in the field. Yet lower canopy layers can contribute up to 50% of the total carbon gain (Long and Drake, 1991; Long *et al.*, 1992). One reason for this may be that within OTCs the upper canopy leaves are exposed to realistic photon fluxes, at least at high sun angles. As a result of the large edge to area ratio of OTCs, lower canopy layers will be subject to less realistic lighting, with increased irradiance due to sidelighting, altered light quality (including ultraviolet wavelengths) and a decreased ratio of direct to diffuse sunlight. Further, physical access to lower canopy layers presents practical problems. Even though leaf gas exchange measurements are non-destructive, access to the lower layers will disturb and may damage the canopy both inside OTCs and FACE plots. However, again this has greater significance when sampling is limited to the small area of OTCs. Understanding of how shade and CO_2 interact, and any change in contribution of the lower canopy versus upper canopy is therefore limited by OTC technology.

Finally, FACE offers advantages in addressing the simple question of how the elevated pCO_2 affects the diurnal course of CO_2 assimilation by leaves. In OTCs and other field enclosures, the interception of direct sunlight and the alteration of the ratio of direct to diffuse light depends on the angle of the sun. Changes to these parameters are most severe at low sun angles, where most radiation must enter through the chamber walls. Again, although sampling of the canopy for "clamp-on" gas exchange measurements is theoretically non-destructive, repeated measurements of the same or adjacent leaves, forced by the small area of OTCs, increases the probability of damage and significant effects on the canopy.

2. *Findings within FACE Experiments*

The FACE experiments with cotton and wheat at Maricopa, and FACE experiments with ryegrass at Eschikon have provided significant information about photosynthetic acclimation. Garcia *et al.* (1998), working with fully irrigated wheat in the Maricopa FACE experiment, provide probably the most comprehensive field study of diurnal courses of leaf photosynthesis and conductance available. They show stimulation at both low photon fluxes at the beginning and end of the day, and in the high photon fluxes of midday. They reveal hysteresis of the stimulation, with a decrease in the early–middle afternoon relative to equivalent photon fluxes in the morning. This depressed stimulation in the afternoon is significantly below that predicted from theory. Averaged over the growing season, the increase in leaf CO_2 assimilation at pCO_2 elevated to 55 Pa was 28% with a 36% decrease in stomatal conductance. The difference in CO_2 assimilation was only lost on completion of grain filling. A similar season-long stimulation of leaf photosynthesis in ryegrass has been observed over successive years in the FACE experiment at Eschikon (Bryant, 1994; Lewis, 1994; Rogers *et al.*, 1998).

Upper canopy leaves of cotton, wheat and ryegrass with high-N fertilisation, showed no acclimation of photosynthesis at any stage of growth, except at completion of grain filling in wheat (Bryant, 1994; Lewis, 1994; Long *et al.*, 1995; Nie *et al.*, 1995b; Osborne *et al.*, 1995, 1998; Rogers *et al.*, 1995, 1998). However, acclimation was found in ryegrass at low levels of nitrogen nutrition, although carbon gain per unit leaf and ground area remained higher. Acclimation at low nitrogen was only apparent in foliage before each harvest. Growth at 60 Pa pCO_2 decreased both the *in vivo* estimate of $V_{c,max}$ and the extracted quantity of Rubisco by 20–40%. After the harvest this difference was lost in the remaining and the new leaves. The result suggests that acclimation was not a direct result of low nitrogen, but resulted from insufficient nitrogen to allow sink expansion feeding back on the photosynthetic apparatus through accumulation and changed fluxes of non-structural carbohydrates in the source leaves (Sheen, 1994; Webber *et al.*, 1994; Drake *et al.*, 1997). When the source : sink ratio was decreased, by removal of part of the canopy, Rubisco levels in the elevated pCO_2 leaves rose to the levels of the controls (Rogers *et al.*, 1995, 1998). Acclimation was selective involving decreased concentrations of the primary carboxylase, Rubisco, rather than a loss of all leaf proteins (Nie *et al.*, 1995b; Rogers *et al.*, 1995). The basis of this response has been shown to be a selective repression of the nuclear encoded gene *rbcS* that encodes part of Rubisco (Nie *et al.*, 1995a,b). These results are both in agreement with the hypothesis that acclimation represents a reallocation of nitrogen away from Rubisco, which, because of low substrate affinity, becomes more efficient under elevated pCO_2 (Webber *et al.*, 1994; Drake *et al.*, 1997). Findings from the first full year of fumigation of loblolly pine in the Duke Forest FACE experiment revealed a slight acclimatory loss of carboxylation capacity at one point in the year only, and no evidence of a loss of the stimulation of photosynthesis at any time (Meyers *et al.*, 1998). The findings in this experiment did show a strong positive dependence of the stimulation of needle photosynthesis on ambient temperature (Meyers *et al.*, 1998). This has been predicted from theory (Long and Drake, 1991) but has rarely been demonstrated within a single field study (Drake *et al.*, 1997).

During grain filling of wheat in the Maricopa FACE experiment, the total photosynthetic C gain of the upper three leaves of the wheat canopy rose by 36% in elevated pCO_2; 39% of this increase was attributed to the second and third leaves from the top of the canopy. In the control plots, the uppermost, i.e. flag, leaves provided 86% of the total, the next layer 15% and the third layer contributed a net deficit of –1%. At elevated pCO_2, the flag leaves contributed 79%, the next layer 19% and the third 2%. Elevated pCO_2 increased the significance of the contribution of the lower leaves to canopy carbon gain. Particularly striking is the fact that, at this stage of development, leaves of the third layer had a negative net C balance over 24 h, but this became positive under elevated pCO_2 (Osborne *et al.*, 1995, 1998). Despite the increase in net C gain, the lower but not upper layers showed acclimation. There was a significant interaction between leaf position in the canopy and pCO_2 for both $V_{c,max}$ and Rubisco content (Osborne *et al.*, 1998). Again, the loss of Rubisco appeared selective, with a second major chloroplast protein (LHC) rising slightly in the same leaves whilst Rubisco declined. The loss of Rubisco corresponded to a decrease in leaf N content and may have resulted from increased demand for N in the grain which constituted a larger sink in elevated pCO_2 (Pinter *et al.*, 1996). The continued stimulation of photosynthesis in lower canopy leaves under elevated pCO_2, despite the loss of Rubisco, parallels the situation in the leaves of ryegrass grown with a low N supply. Functionally, it may similarly be explained by the fact that Rubisco was in excess of the quantities that could limit photosynthesis at elevated pCO_2 (Osborne *et al.*, 1998; Rogers *et al.*, 1998).

Studies of carbohydrates and of secondary metabolites underline the requirement for sufficient material to allow frequent destructive sampling. Both show an important interaction between pCO_2 treatment, time of day and developmental stage. Leaf carbohydrates at anthesis of wheat showed large differences between control and elevated pCO_2 in the late afternoon but none in the morning. Similarly, differences apparent at anthesis were absent during grain-filling (Nie *et al.*, 1995a). Anti-oxidant levels in leaves also showed large daily and seasonal dynamics (Badiani *et al.*, 1996). In both cases very different conclusions might have been drawn had the amount of material been only sufficient for two or three sets of samples. Fischer *et al.* (1997) similarly showed that differences in leaf carbohydrate concentrations in ryegrass in response to pCO_2 elevated to 60 Pa were highly dependent on growth stage and could vary over a few days.

3. How Has FACE Influenced Understanding?

In summary, FACE has discriminated between different suggestions from OTC and laboratory studies of how wheat (the world's major food crop) and ryegrass respond to elevated pCO_2. FACE experiments have demonstrated a persistent increase in leaf photosynthetic C gain at all times of day and in different canopy layers, throughout the life of annual crops and throughout the seasons in pasture and forest species (Bryant, 1994; Lewis, 1994; Garcia *et al.*, 1998; Meyers *et al.*, 1998; Osborne *et al.*, 1998; Rogers *et al.*, 1998). FACE has demonstrated, across four large series of experiments, an almost complete lack of evidence for any loss of the initial stimulation of leaf photosynthesis by elevated pCO_2 in the long term. This is in sharp contrast to the prevailing expectation of a long-term decline (Oechel *et al.*, 1994). However, whether this persistent increase is a result of fumigation method or of contrasts in ecosystems awaits comparisons within the same system. These experiments have also confirmed the role of N in photosynthetic acclimation under elevated pCO_2 and have provided new information on the dynamics of acclimation at the level of gene expression. Whilst much of this might have been inferred by combining information from different laboratory and OTC studies, FACE has allowed the simultaneous investigation of the full range of factors contributing to acclimation. These include position in the canopy and whole canopy exchanges, the dynamics of carbohydrate concentrations and the levels of mRNAs coding for key polypeptides of the photosynthetic apparatus. FACE has also provided the first clear information of the absolute contributions of upper and lower canopy leaves to increase carbon gain in undisturbed field canopies. However, FACE experiments are too few to determine whether these patterns can be generalised.

D. Effects on Roots and Soil

Annual crops may show a 1.5 m lateral spread of roots from an individual. For perennial plants, and especially trees, the spread can be much greater (Russell, 1961). In light soils, lateral movement of water and therefore nutrients can be considerable (Russell, 1961). Extraction of root systems by soil cores cannot be made without significant disturbance of the adjacent areas. Thus, if the limited size of OTCs and other enclosures present problems for studies of the leaf canopy, these problems are yet greater for studies of roots and soils.

1. Limitations of OTCs and the Theoretical Advantages of FACE

Given the spread of plant root systems, many species will have the potential to exploit soil resources outside the area of ground covered by an open-top chamber. For a species that responds positively to elevated pCO_2, this will favour it in competing with plants in the unenriched environment outside the chamber. Similarly, changes in the rates of removal of water or nutrients may be offset by lateral movement. These problems may be minimised by insertion of an impermeable barrier, but this creates further problems. First, physical impedance of root distribution can cause an acclimatory decrease in photosynthetic capacity (Masle *et al.*, 1990; Arp, 1991). Secondly, inserting the barrier requires that much of the soil system will be disturbed. The roots will therefore grow in very disturbed soil and will be restricted, imposing an additional "chamber" effect on the experiment. Laboratory studies suggest that alteration of allocation patterns by elevated pCO_2 is dynamic. Thus, investigation of effects on root growth and patterns of allocation require regular sampling.

Direct measurement of allocation to roots in the field requires physical extraction of the root system, although changes can be partly inferred from non-destructive rhizotron observations or from ingrowth bag measurements. Extraction of even two or three cores within a single OTC is likely to affect the root systems of a large proportion of the plants, affecting drainage and causing physical disturbance to much of the canopy. For soil C accumulation, large numbers of cores would be required to gain sufficient statistical power to resolve an effect of elevated pCO_2 (Hungate *et al.*, 1996). Consequently, repeated extraction of cores will be impractical. FACE has provided the increased scale needed to assess possible increased rates of soil exploration and increased soil C accumulation by frequent and extensive extraction of soil cores. It also provides sufficient plot size to allow unambiguous interpretation of data from mini-rhizotrons.

Alteration of soil moisture through altered coupling and chamber interception of precipitation can cause a large chamber effect on the soil (see section V.B). A number of changes in soil microbial activity and root growth have been attributed to increased conservation of soil moisture under elevated pCO_2. Whilst OTCs may identify these effects, accurate quantification will only be possible where coupling is unaltered.

As a result of the large net flow of air out of an OTC or other enclosure system, pressure within the OTC has to exceed atmospheric pressure and air is therefore forced into the soil. The varied and fluctuating patterns of air flow within chambers mean that the exact pressure difference at any one point cannot easily be defined nor is it necessarily constant. Accurate measurement of soil CO_2 flux, resulting from root and microbial activity, or fluxes of any trace gases are therefore precluded.

2. *Findings within FACE Experiments*

In the FACE experiments with cotton and wheat at Maricopa, soil cores were removed at regular intervals from the replicate plots. There was sufficient space to extract these cores without interfering with other measurements of the crops. Both crops showed significant and large increases in belowground biomass, amounts of fine root and, most importantly, a greatly increased initial rate of soil exploration (Rogers *et al.*, 1992; Prior *et al.*, 1993, 1994a,b, 1995, 1997; Rogers and Dahlman, 1993; Wechsung *et al.*, 1995). In wheat, root mass was very much higher at elevated pCO_2 in the early stages of growth but the difference declined with development. At the three-leaf stage, root mass was 36% higher, declining to 19% at harvest (Wechsung *et al.*, 1995). Roots of the wheat plants in elevated pCO_2 explored a significantly greater proportion of the soil profile at the early stages of growth, but by maturity the root system of the crop grown under elevated pCO_2 was not found at significantly greater depths than controls (Wechsung *et al.*, 1995). In the Maricopa FACE experiments (sections III.B.1–2) the CO_2 used for enrichment was from petrochemical sources and was depleted in both radiocarbon (^{14}C content 0% modern carbon (pmC)) and ^{13}C ($\delta^{13}C$ approximately –36‰) relative to background air. The enrichment CO_2 provided a strong isotopic tracer for evaluation of C incorporation into soil. In the FACE experiment with cotton, the soil stable-C isotope ratio and mass-balance calculations indicated that 10% of the organic carbon in the FACE plot soil derived from the 3-year elevated pCO_2 treatment (Leavitt *et al.*, 1994). Greater increases in soil C and belowground biomass have been suggested from many previous studies, as have the large changes in allocation towards roots. These FACE studies differ in showing the dynamics of this development and avoiding the uncertainties resulting from the limited sampling of soil and the disturbance associated with OTC studies.

Increases in root production and growth were accompanied by increased efflux of soil CO_2 and other trace gases, coupled with increased soil microbial and microfauna population sizes (Nakayama *et al.*, 1994; Runion *et al.*, 1994). However, whilst cotton residues from FACE showed slower rates of soil decomposition, the decomposability of wheat produced in FACE at Maricopa was apparently unchanged (Wood *et al.*, 1994; Akin *et al.*, 1995b).

Jongen *et al.* (1995) showed highly significant increases in root growth of both clover and ryegrass in ingrowth cores at the Eschikon FACE experiment during the first year of growth in elevated pCO_2. Subsequent analyses of soil cores have also shown that, after 3 years, there has been a 26–40% increase in root mass of clover and a 45–108% increase in root mass of ryegrass (Hebeisen *et al.*, 1997b). Although large increases in photosynthesis have been observed at elevated pCO_2 in ryegrass at both high and low nitrogen, shoot production was only observed to increase at high N, and then only by 7% (Hebeisen *et al.*, 1997b). This apparent mismatch is now explained by the significant increase in root production and allocation observed in this treatment. Two further experiments within the Eschikon FACE experiment have imported soils that have developed under C_4 species providing a distinctive ^{13}C signal within the organic matter of the soil. When a C_3 species is planted on such a soil, the amount of carbon that the species adds to the soil can be determined from the shift in the ^{13}C signal. Nitschelm *et al.* (1997) planted clover on such a soil and showed that an addition of 200 g C m^{-2} at ambient pCO_2 increased significantly to 300 g C m^{-2} at an elevated pCO_2 of 60 Pa, after 1 year. Ineson *et al.* (1996) planted birch seedlings (*Betula pendula*) and using ^{13}C showed that, after one growing season, elevated pCO_2 had increased C in root matter by 69% and soil C resulting from root exudation or turnover by 150%. Shoot biomass for the same plants did not increase in elevated pCO_2.

Increased microbial activity is associated with these increases in production. Zanetti *et al.* (1996) showed with ^{15}N dilution that *Trifolium repens* in elevated pCO_2 using FACE showed an increase in symbiotically fixed N of 6.2–8.2 g m^{-2} $year^{-1}$ over 3 years. In both pure and the clover–ryegrass mixture the additional N in the increased biomass was almost entirely accounted for by the increased symbiotic N_2 fixation (Zanetti and Hartwig, 1997). Numbers of heterotrophic bacteria, excepting *Rhizobium,* were not found to increase in the soil under the elevated pCO_2 treatment (Schortemeyer *et al.*, 1996). However, activity was increased as indicated by increased rates of soil CO_2 efflux and other trace gases (Ineson *et al.*, 1997).

3. How Has FACE Influenced Understanding?

FACE studies have provided the first evidence that there are highly significant increases in both soil CO_2 efflux and the efflux of other trace gases from ecosystems growing in elevated pCO_2. Whilst these were inferred from OTC studies, the forced movement of air under pressure from the chamber into the soil prevented any certainty of estimates, not only with respect to the magnitude but also the direction of flux. FACE eliminates the pressure difference and provides scale, allowing estimates that are unlikely to be affected by exchange with the soil atmosphere outside of the treated area. Increases in root and soil C have frequently been suggested from OTC studies or inferred from observed increases in C assimilation without increases in shoot primary production. However, the uncertainties of whether differences persist throughout the growing season and in the long term result from the difficulty of removing large numbers of soil cores in the limited space of OTCs. Disturbance of the soil profile in erecting enclosures and some miniFACE systems with buried pipes add further uncertainties. As noted by Hungate *et al.* (1996), new approaches would be necessary to provide adequate statistical power to detect increase in soil C. FACE experiments appear to have started to meet this requirement. Much of the observed impact of elevated pCO_2 has been associated with improved soil moisture (e.g. Field *et al.*, 1997; Owensby, *et al.*, 1997). As noted earlier, the relative magnitude of this effect will depend on atmospheric coupling that is altered in a very complex manner by OTCs. Consequently, OTCs cannot provide a reliable method of determining the magnitude of changes in rooting mass, soil microbial activity or soil trace grass effluxes, but only their direction of change. The FACE studies now provide clear, statistically significant evidence for increases in rooting mass and soil C with both the arable crops wheat and cotton, and with the pasture crops ryegrass and clover. FACE studies also reveal greatly accelerated growth and faster rates of soil exploration in the early stages of crop development. Most significantly, by use of ^{13}C, very large increases in soil C input with growth of a range of plants under elevated pCO_2 have been revealed. These suggest surprisingly high additional rates of carbon accumulation in soils under elevated pCO_2. It will be critical to extend these analyses to assess whether these higher rates are maintained into the long term. Biogenic N_2-fixation is similarly increased by surprising amounts, as suggested previously by a number of OTC studies. However, FACE has shown that, in the longer term for pastures, this accounts for all of the additional N required to sustain a continued increase in productivity under elevated pCO_2.

E. Interactions

The scale of FACE experiments and the absence of a physical barrier, allows a more realistic assessment of the interactions of pCO_2 with pests and pathogens, but this potential has not been explored. FACE can similarly provide the space for the study of a number of plant species interactions. However, the physical isolation of FACE plots, the "island effect" (section IV.D) places some constraints upon the interpretation of pest/pathogen interactions with treatment (McLeod, 1993).

1. Limitations of OTCs and the Theoretical Advantages of FACE

In the FACE experiment at Eschikon (Hebeisen *et al.*, 1997b) and the tall grass prairie experiment at Cedar Creek, Minnesota (Table 1; P. Reich, personal communication), the substantial treated plot area has been subdivided into relatively small plots. This loses the scale advantage of FACE, since similar areas could be enclosed in OTCs or planted in miniFACE facilities. The following explains why this use of large FACE systems still retains advantages.

A FACE plot of 20 m diameter could enclose 85 plots of 2 m^2 separated from each other by a minimum guard distance of 0.5 m and would be adequate for short vegetation. Such experiments could equally be accommodated into a large number of OTCs, but at some 3.6 times the running costs of FACE (Table 3; Kimball *et al.*, 1992). Another potential advantage would be that the 85 potential treatments or treatment combinations would be truly blocked within the FACE plot. Any undetected differences between the FACE plots would therefore also be blocked. If the treatments were arranged in blocks of OTCs, an undetected difference in one OTC would be confounded statistically with the treatment. In the pasture experiments at Eschikon, these advantages of FACE have been used to assess competition between some 15 species and a similar number of populations, and also to study the performance of ryegrass and clover, in isolation and in combination under two N and two cutting regimes. In a recent FACE experiment "BioCon" (Biodiversity, CO_2 and nitrogen) at the University of Minnesota, between 1, 4, 9 or 16 species, randomly selected from 16 common species in local grasslands and prairies have been planted into 61 2 m × 2 m plots within three replicated elevated pCO_2 and control plots to assess multidimensional species interactions (G. Hendrey, personal communication).

2. *Findings within FACE Experiments*

In the FACE experiment at Eschikon, mixed stands of ryegrass and clover were planted in sub-plots, alongside sub-plots of the monocultures (section III.B.3). A greater responsiveness of clover in the monocultures was observed in the first year, suggesting that this species was favoured by rising pCO_2. This expectation was realised in the third year when highly significant increases in the proportion of clover in the mixed plots were found. This was especially pronounced in the nested plots receiving low levels of N fertilisation and in the most frequently defoliated plots (Hebeisen *et al.*, 1997b). In further sub-plots within the FACE plots at Eschikon, interspecific and intraspecific variability in the responsiveness to pCO_2 elevated to 60 Pa was determined for 9–14 different genotypes of each of 12 perennial species from fertile permanent grassland. The species were grouped according to their functional types: seven grasses, three non-legume dicotyledons and two legumes.

Yield based on taking three cuts per year showed highly significant differences with respect to pCO_2 between the three functional types. The two legumes showed the strongest and the seven grasses the weakest response. The response of the legumes was strongest in the first year whilst the non-leguminous dicotyledons showed an increase in response into the third year. The interspecific differences suggest likely changes in the species composition of fertile temperate grassland with rising pCO_2. Owing to the temporal differences in the responsiveness to elevated pCO_2 among species, complex effects of elevated pCO_2 on competitive interactions in mixed swards must be expected. The existence of genotypic variability in the responsiveness to elevated pCO_2, on which selection could act, was not found in the Eschikon FACE experimental conditions (Luscher *et al.*, 1998).

3. *How Has FACE Influenced Understanding?*

Although interactions between species have formed a small amount of the output from FACE experiments so far, the findings have been nonetheless of considerable importance. It has long been suspected that elevated pCO_2 would favour legume species relative to associated grasses. However, these experiments have been able to couple simultaneous measurements of biological N fixation, soil N changes, photosynthesis and production, with proof that competition and survival are modified over a period of years to favour legumes. This is the first evidence of actual species composition changes under open-air conditions. Although it confirms suggestions from enclosed chamber studies, removal of the modifications of light, temperature and rainfall imposed by enclosures and other factors likely to alter competitive interactions make these findings a significant advance.

VI. CONCLUSIONS

Free-air carbon dioxide enrichment (FACE) systems have provided a means to study the effects of elevated pCO_2 on vegetation and many other ecosystem components in large-scale outdoor plots (> 20 m diameter). Control of pCO_2 in large-scale FACE experiments has now been developed to an extent where long-term means (over 6 months or 1 year) achieve a similar percentage of the desired target pCO_2 as achieved with large-scale closed-chamber facilities (section IV). Typically a BNL FACE facility can achieve 1 min mean pCO_2 to within ± 10% of the set point of 55 Pa for > 90% of the time and ± 20% for > 99% averaged over two seasons' experiments.

FACE systems use large amounts of CO_2 gas. Typical CO_2 usage on 23 m diameter FACE arrays at Maricopa, Arizona, from May to September was 1400 kg per day for 12 h of fumigation to a target of 55 Pa (section IV.E). FACE systems also require a considerable amount of equipment to establish and large quantities of CO_2 annually. Despite high total costs, when fully utilised and expressed on a cost per square metre of usable experimental area, or per research scientist or subject of study, the FACE method is more cost effective than alternative methods of pCO_2 enrichment.

The FACE method avoids many modifications to the microclimate that occur in open-top or closed chamber studies where "chamber effects" on temperature, irradiance and water relations may be greater than the effects of doubling pCO_2. The aerial microclimate in chambers is typically duller, warmer and drier than ambient conditions. The forced airflow inside open-top and closed chambers makes atmospheric coupling very different from ambient conditions. It is not known how this may affect the increased stomatal resistance of vegetation to water vapour loss in elevated pCO_2. Although FACE provides a means to minimise alteration of coupling, this may only be true in continuous vegetation where there are no "edge" effects. Not all FACE experiments have continuous vegetation and small miniFACE systems (< 1 m diameter) may be dominated by "edge" effects.

The FACE method has been used for two series of experiments on cotton and wheat, on the grassland species, ryegrass and clover, and in a mature canopy of loblolly pine. All of these studies have demonstrated advantages of the large scale of FACE plots. Simultaneous investigations and sampling by many research groups (national and international) and their exchange of observations allows a very complete analysis and facilitates a synergism between research areas that could not be gained at a smaller scale. It also allows comparisons of different methods within a plot and a degree of cross-referencing with other studies of the same material only possible at this scale. The FACE-cotton experiments involved 47 investigators from 23 research groups while the FACE-wheat experiments involved over 50 scientists from 28 research groups from 8 countries.

Careful site management and planning has demonstrated that site trampling can be adequately controlled even with large numbers of researchers. "Edge effects" are also minimised in FACE plots but many of the advantages of scale are not achieved by miniature FACE systems with a plot size of 1 m or less (section III.B.5).

A number of designs for such small FACE systems with 0.15–1.0 m plots (often termed miniFACE) have been described. The information on gas release concentration and on spatial distribution of concentration (section IV.B.1) suggests that there may be large concentration gradients and large fluctuations in concentration within miniFACE plots. There is now evidence that rapid fluctuations in pCO_2 typical of some enrichment systems can affect plant physiological responses to elevated pCO_2. Enhanced local mixing devices to improve dispersion, reduce concentration fluctuations and which can achieve a theoretical 40% increase in gas utilisation efficiency in FACE have been described. A new FACE design incorporating a dual ring of release pipes with an outer ring of pipes blowing air into the wind in order to reduce CO_2 use has not been adequately demonstrated and it is considered unlikely to improve gas use efficiency.

Studies on crop production under elevated pCO_2 using FACE have confirmed some earlier observations made with OTC studies. They have also revealed new information and given greater quantitative certainty to crop yields under elevated pCO_2 at an agronomically meaningful scale (section V.A). Cotton yield under FACE increased by about 40% at 55 Pa, confirming predictions from OTC studies. However, the lower wheat yield increase of 8–12% under FACE compares with an equivalent 20% predicted from OTC studies (see section V.A.3) demonstrating the value of using an alternative technique. Prior to the FACE experiments it had also been assumed when predicting future food supplies under climate change that elevated pCO_2 would increase yields less under stress conditions. However, the percentage increase in wheat yield under FACE and full irrigation was 8–12% compared to 21–25% under FACE and drought treatment. Although FACE has therefore provided important new information, it must be remembered that no single crop or vegetation type has yet been examined at more than one location using FACE. Future replication of FACE studies will be important if the range of site differences in wheat responses observed in OTC studies are representative.

FACE experiments have contributed important results to the debate on decreased photosynthetic potential under elevated pCO_2 and its significance to net primary production. Results from the wheat experiments at Maricopa and the ryegrass experiment at Eschikon (section V.C.2) have both demonstrated acclimation of photosynthesis under conditions of low N, but no loss of enhancement of photosynthesis. Results confirm the hypothesis that acclimation represents a reallocation of N away from Rubisco, which is more efficient under elevated pCO_2.

FACE experiments at Maricopa have detected an increase in canopy temperature in elevated pCO_2 plots not only during the day, which was attributed to stomatal closure and reduced evapotranspiration, but also during the night. The night-time temperature increase was attributed to the air fans circulating warmer air from greater heights during frequent inversion conditions (see section IV.C). This did not occur on ambient plots without blowers. This effect of blowers constitutes a "chamber effect" of FACE, which was confirmed by studies using ambient plots both with and without blowers. This "blower effect", together with the "island effect" (section IV.D) of FACE plot isolation and the influence of plot isolation on water vapour flux (and also on biological effects on pests and pathogens) is one area of FACE experimentation that warrants further analysis. The FACE experiments at Maricopa provide convincing data, in the form of energy balance measurements and false-colour thermal images, that reduced transpiration caused by elevated pCO_2 will lead to increases in regional surface temperatures of 1–2°C. This potential feedback of vegetation on global climate is similar in magnitude to that predicted to arise directly from the increased atmospheric concentration of greenhouse gases. Of further significance are the early observations from the Duke Forest FACE experiment that suggest no direct response of needle transpiration of plant water relations with implications for the role of boreal forest in such warming.

The FACE experiments at Maricopa have provided clear evidence of greatly increased belowground biomass of cotton and wheat under elevated pCO_2, large increases in amounts of fine root and a greatly increased initial rate of soil exploration. Increased soil CO_2 release and trace gas effluxes coupled with increased soil microbial populations were confirmed in FACE experiments, providing more confidence in the applicability of results from OTCs. The use of carbon-isotype dynamics in combination with FACE has proved a useful tool for assessment of soil C accumulation.

The new scientific findings from FACE experiments noted above, in addition to the confirmation of earlier knowledge from other techniques, demonstrate the value of FACE methodology. The potential for application of the technique to a range of crops and ecosystems has now been amply demonstrated to be both feasible and effective. There are plans to apply FACE in studies of a range of further ecosystems worldwide and the method forms an important component in the International Geosphere–Biosphere Programme (IGBP) core project on Global Change and Terrestrial Ecosystems (GCTE, 1992). On past experience, outlined in this chapter, the information obtained has the potential to improve our understanding greatly of the impacts of global atmospheric change on terrestrial ecosystems.

ACKNOWLEDGMENTS

The authors thank all those who have provided information for use in this chapter. The financial support provided by the NERC and BBSRC through the joint initiative on CO_2 facilities of the TIGER (Terrestrial Initiative in Global Environment Research) and BAGEC (Biological Adaptation to Global Environment Change) programmes, award number T94/CO_2/4 is gratefully acknowledged.

REFERENCES

Akin, D.E., Kimball, B.A., Windham, W.R., Pinter, P.J., Wall, G.W., Garcia, R.L., Lamorte, R.L. and Morrison, W.H. (1995a). Effect of Free-Air CO_2 Enrichment (FACE) on forage quality of wheat. *Animal Feed Science and Technology* **53**, 29–43.

Akin, D.E., Rigsby, L.L., Gamble, G.R., Morrison, W.H., Kimball, B.A., Pinter, P.J., Wall, G.W., Garcia, R.L. and Lamorte, R.L. (1995b). Biodegradation of plant-cell walls, wall carbohydrates, and wall aromatics in wheat grown in ambient or enriched CO_2 concentrations. *Journal of the Science of Food and Agriculture* **67**, 399–406.

Allen, L.H.J. (1992). Free-Air CO_2 enrichment field experiments—a historical overview. *Critical Reviews in Plant Sciences* **11**, 121–134.

Allen, L.H.J., Drake, B.G., Rogers, H.H. and Shinn, J.H. (1992). Field techniques for exposure of plants and ecosystems to elevated CO_2 and other trace gases. In: *FACE: Free-Air CO_2 Enrichment for Plant Research in the Field* (Ed. by G.R. Hendrey). CRC Press Inc, Boca Raton, FL.

Amthor, J.S. (1994). Scaling CO_2-photosynthesis relationships from the leaf to the canopy. *Photosynthesis Research* **39**, 321–350.

Apsley, D.D. (1991). *Design Study for an Open-air CO_2 Fumigation Facility.* Report No. TEC/L/0080/TAN91, May 1991. National Power Technology and Environmental Centre, Leatherhead, UK.

Arp, W.J. (1991). Effects of source-sink relations on photosynthetic acclimation to elevated CO_2. *Plant Cell Environ.* **14**, 869–875.

Badiani, M., Paolacci, A.R., Miglietta, F., Kimball, B.A., Pinter, P.J., Garcia, R.L., Hunsaker, D.J., LaMorte, R.L. and Wall, G.W. (1996). Seasonal variations of antioxidants in wheat (*Triticum aestivum*) leaves grown under field conditions. *Australian Journal of Plant Physiology* **23**, 687–698.

Bhattacharya, N.C., Radin, J.W., Kimball, B.A., Mauney, J.R., Hendrey, G.R., Nagy, J., Lewin, K.F. and Ponce, D.C. (1994). Leaf water relations of cotton in a Free-Air CO_2 Enriched environment. *Agricultural and Forest Meteorology* **70**, 171–182.

Blum, H. (1993). The response of CO_2-related processes in grassland ecosystems in a three-year FACE study. In: *Design and Execution of Experiments on CO_2 Enrichment. Proceedings of a Workshop held at Weidenberg, Germany, 26–30 October 1992* (Ed. by E.D. Schulze and H.A. Mooney), pp. 367–370. Ecosystems Research Report 6, Commission of the European Communities/E Guyot SA, Brussels.

Blum, H., Hendrey, G. and Nosberger, J. (1997). Effects of elevated CO_2, N fertilization, and cutting regime on the production and quality of *Lolium perenne* L. shoot necromass. *Acta Oecologica—International Journal of Ecology* **18**, 291–295.

Bryant, J.B. (1994). *The Photosynthetic Acclimation of* Lolium perenne *Growing in a Free-Air CO_2 Enrichment (FACE) System.* M.Sc. thesis, University of Essex, Colchester.

Bucher, J.B., Tarjan, D.P., Siegwolf, R.T.W., Saurer, M., Blum, H. and Hendrey, G.R. (1998). Growth of a deciduous tree seedling community in response to elevated CO_2 and nutrient supply. *Chemosphere* **36**, 777–782.

Cardon, Z.G., Berry, J.A. and Woodrow, I.E. (1994). Dependence of the extent and direction of average stomatal response in *Zea mays* L. and *Phaseolus vulgaris* L. on the frequency of fluctuations in environmental stimuli. *Plant Physiology* **105**, 1007–1013.

Cardon, Z.G., Berry, J.A. and Woodrow, I.E. (1995). Fluctuating [CO_2] drives species-specific changes in water use efficiency. *Journal of Biogeography* **22**, 203–208.

Curtis, P.S., and Wang, X.Z. (1998). A meta-analysis of elevated CO_2 effects on woody plant mass, form, and physiology. *Oecologia* **113**, 299–313.

Day, F.P., Weber, E.P., Hinkle, C.R. and Drake, B.G. (1996). Effects of elevated atmospheric CO_2 on fine-root length and distribution in an oak-palmetto scrub ecosystem in central Florida. *Global Change Biology* **2**, 143–148.

Drake, B.G. (1996). Long-term elevated CO_2 exposure in a Chesapeake Bay wetland: Ecosystem gas exchange, primary production, and tissue nitrogen. In: *Carbon Dioxide and Terrestrial Ecosystems* (Ed. by G.W. Koch and H.A. Mooney), pp. 197–214. Academic Press, San Diego.

Drake, B.G., Leadley, P.W., Arp, W.J., Nassiry, D. and Curtis, P.S. (1989). An open top chamber for field studies of elevated atmospheric CO_2 concentration on saltmarsh vegetation. *Functional Ecology* **3**, 363–371.

Drake, B.G., Gonzalez-Meler, M. and Long, S.P. (1997). More efficient plants: A consequence of rising atmospheric CO_2? *Annual Review of Plant Physiology and Plant Molecular Biology* **48**, 609–639.

Dugas, W.A. and Pinter, P.J. (1994). Introduction to the Free-Air Carbon Dioxide Enrichment (FACE) cotton project. *Agricultural and Forest Meteorology* **70**, 1–2.

Dugas, W.A., Heuer, M.L., Hunsaker, D., Kimball, B.A., Lewin, K.F., Nagy, J. and Johnson, M. (1994). Sap flow measurements of transpiration from cotton grown under ambient and enriched CO_2 concentrations. *Agricultural and Forest Meteorology* **70**, 231–245.

Ellsworth, D.S. (1998). Are CO_2 enrichment in a maturing pine forest: Are CO_2 exchange and water status in the canopy affected? *Plant, Cell and Environment* (in press).

Ellsworth, D.S., Oren, R., Huang, C., Phillips, N., and Hendrey, G.R. (1995). Leaf and canopy responses to elevated CO_2 in a pine forest under free-air CO_2 enrichment. *Oecologia* **104**, 139–146.

Estiarte, M., Penuelas, J., Kimball, B.A., Idso, S.B., Lamorte, R.L., Pinter, P.J., Wall, G.W. and Garcia, R.L. (1994). Elevated CO_2 effects on stomatal density of wheat and sour orange trees. *Journal of Experimental Botany* **45**, 1665–1668.

Farrow, T., Nagy, J., Flewin, K. and Hendrey, G.R. (1992). *Signal Delay and Time Constant of the CO_2 Concentration Measurement System of the FACE Experiment at Maricopa, AZ.* BNL-48282. Brookhaven National Laboratory, Upton, NY 11973, USA.

Field, C.B., Lund, C.P., Chiariello, N.R. and Mortimer, B.E. (1997). CO_2 effects on the water budget of grassland microcosm communities. *Global Change Biology* **3**, 197–206.

Fischer, B.U., Frehner, M., Hebeisen, T., Zanetti, S., Stadelmann, F., Luscher, A., Hartwig, U.A., Hendrey, G.R., Blum, H. and Nosberger, J. (1997). Source-sink relations in *Lolium perenne* L. as reflected by carbohydrate concentrations in leaves and pseudo-stems during regrowth in a free air carbon dioxide enrichment (FACE) experiment. *Plant Cell Environ.* **20**, 945–952.

Fredeen, A.L., Randerson, J.T., Holbrook, N.M. and Field, C.B. (1997). Elevated atmospheric CO_2 increases water availability in a water-limited grassland ecosystem. *Journal of the American Water Resources Association* **33**, 1033–1039.

Frehner, M., Luscher, A., Hebeisen, T., Zanetti, S., Schubiger, F. and Scalet, M. (1997). Effects of elevated partial pressure of carbon dioxide and season of the year on forage quality and cyanide concentration of *Trifolium repens* L. from a FACE experiment. *Acta Oecologica—International Journal of Ecology* **18**, 297–304.

Fuhrer, J. (1994). Effects of ozone on managed pasture. 1. Effects of open-top chambers on microclimate, ozone flux, and plant-growth. *Environmental Pollution* **86**, 297–305.

Garcia, R.L., Long, S.P., Wall, G.W., Osborne, C.P., Kimball, B.A., Nie, G.Y., Pinter, P.J., Lamorte, R.L. and Wechsung, F. (1998). Photosynthesis and conductance of spring wheat leaves: Field response to continuous free-air atmospheric CO_2 enrichment. *Plant Cell Environ.* **21** (in press).

GCTE (1992). *Global Change and Terrestrial Ecosystems. The Operational Plan.* IGBP Report No. 21. International Council of Scientific Unions, Stockholm.

Ghannoum, O., von Caemmerer, S., Barlow, E.W.R. and Conroy, J.P. (1997). The effect of CO_2 enrichment and irradiance on the growth, morphology and gas exchange of a C_3 (*Panicum laxum*) and a C_4 (*Panicum antidotale*) grass. *Australian Journal of Plant Physiology* **24**, 227–237.

Grant, R.F., Garcia, R.L., Pinter, P.J., Hunsaker, D., Wall, G.W., Kimball, B.A. and Lamorte, R.L. (1995). Interaction between atmospheric CO_2 concentration and water-deficit on gas-exchange and crop growth—testing of ECOSYS with data from the free-air CO_2 enrichment (FACE) experiment. *Global Change Biology* **1**, 443–454.

Grossman, S., Kartschall, T., Kimball, B.A., Hunsaker, D.J., Lamorte, R.L., Garcia, R.L., Wall, G.W. and Pinter, P.J. (1995). Simulated responses of energy and water fluxes to ambient atmosphere and free-air carbon dioxide enrichment in wheat. *Journal of Biogeography* **22**, 601–609.

Gunderson, C.A. and Wullschleger, S.D. (1994). Photosynthetic acclimation in trees to rising atmospheric CO_2—a broader perspective. *Photosynthesis Research* **39**, 369–388.

Heagle, A.S., Philbeck, R.B., Ferrell, R.E. and Heck, W.W. (1989). Design and performance of a large field exposure chamber to measure effects of air-quality on plants. *Journal of Environmental Quality* **18**, 361–368.

Heath, J. and Kersteins, G. (1996). Effects of elevated CO_2 on leaf gas exchange in beech and oak at two levels of nutrient supply: Consequences for sensitivity to drought in beech. *Plant Cell Environ.* **20**, 57–67.

Hebeisen, T., Luscher, A. and Nosberger, J. (1997a). Effects of elevated atmospheric CO_2 and nitrogen fertilisation on yield of *Trifolium repens* and *Lolium perenne. Acta Oecologica—International Journal of Ecology* **18**, 277–284.

Hebeisen, T., Luscher, A., Zanetti, S., Fischer, B.U., Hartwig, U.A., Frehner, M., Hendrey, G.R., Blum, H. and Nosberger, J. (1997b). Growth response of *Trifolium repens* L and *Lolium perenne* L as monocultures and bi-species mixture to free air CO_2 enrichment and management. *Global Change Biology* **3**, 149–160.

Hendrey, G.R. (1993). *FACE: Free-Air CO_2 Enrichment for Plant Research in the Field,* CRC Press, Boca Raton, FL 33431, USA, 308 pp.

Hendrey, G.R., Lewin, K.F. and Nagy, J. (1993). Free-air carbon dioxide enrichment —development, progress, results. *Vegetatio* **104**, 17–31.

Hendrey, G.R., Lewin, K.F. and Nagy, J. (1994). *Brookhaven National Laboratory Free Air Carbon Dioxide Enrichment Forest Prototype*. Interim Report, August 22, 1994. Brookhaven National Laboratory, Upton, NY.

Hendrey, G.R., Long, S.P., Baker, N.R. and McKee, I.F. (1995). Response of leaf photosynthesis to short-term fluctuations in atmospheric carbon dioxide. In: *Photosynthesis: From Light to Biosphere* (Ed. by P. Mathis), Vol. 5, pp. 965–968. Kluwer Academic Publishers, Dordrecht, The Netherlands.

Hendrey, G.R., Long, S.P., McKee, I.F. and Baker, N.R. (1997). Can photosynthesis respond to short-term fluctuations in atmospheric carbon dioxide? *Photosynthesis Research* **51**, 179–184.

Hendrey, G.R., Ellsworth, D.S., Lewin, K.F. and Nagy, J. (1998). A free-air enrichment system for exposing tall forest vegetation to elevated atmospheric CO_2. *Global Change Biology* **4** (in press).

Hertstein, U., Fangmeier, A. and Jager, H.J. (1996). ESPACE-wheat (European Stress Physiology and Climate Experiment—Project 1: Wheat): Objectives, general approach, and first results. *Journal of Applied Botany—Angewandte Botanik* **70**, 172–180.

Hileman, D.R., Ghosh, P.P., Bhattacharya, N.C., Biswas, P.K., Allen, L.H., Peresta, G. and Kimball, B.A. (1992). A comparison of the uniformity of an elevated CO_2 environment in three different types of open-top chambers. *Critical Reviews in Plant Sciences* **11**, 195–202.

Hileman, D.R., Huluka, G., Kenjige, P.K., Sinha, N., Bhattacharya, N.C., Biswas, P.K., Lewin, K.F., Nagy, J. and Hendrey, G.R. (1994). Canopy photosynthesis and transpiration of field-grown cotton exposed to free-air CO_2 enrichment (FACE) and differential irrigation. *Agricultural and Forest Meteorology* **70**, 189–207.

Hungate, B.A., Jackson, R.B., Field, C.B. and Chapin, F.S. (1996). Detecting changes in soil carbon in CO_2 enrichment experiments. *Plant and Soil* **187**, 135–145.

Hungate, B.A., Holland, E.A., Jackson, R.B., Chapin, F.S., Mooney, H.A. and Field, C.B. (1997). The fate of carbon in grasslands under carbon dioxide enrichment. *Nature* **388**, 576–579.

Hunsaker, D.J., Kimball, B.A., Pinter, P.J., Lamorte, R.L. and Wall, G.W. (1996). Carbon dioxide enrichment and irrigation effects on wheat evapotranspiration and water-use efficiency. *Transactions of the ASAE* **39**, 1345–1355.

Idso, S.B., Kimball, B.A., Wall, G.W., Garcia, R.L., Lamorte, R., Pinter, P.J., Mauney, J.R., Hendrey, G.R., Lewin, K. and Nagy, J. (1994). Effects of free-air CO_2 enrichment on the light response curve of net photosynthesis in cotton leaves. *Agricultural and Forest Meteorology* **70**, 183–188.

Ineson, P., Cotrufo, M.F., Bol, R., Harkness, D.D. and Blum, H. (1996). Quantification of soil carbon inputs under elevated CO_2: C_3 plants in a C_4 soil. *Plant and Soil* **187**, 345–350.

Ineson, P., Coward, P.A. and Hartwig, U.A. (1997). Soil gas fluxes of N_2O, CH_4 and CO_2 beneath *Lolium perenne* under elevated CO_2: the Swiss FACE experiment. *Plant and Soil* **198**, 89–95.

Isebrands, J.G., Hendrey, G., Coleman, M., Lewin, K., Nagy, J., Sober, J., Kruger, E., Pregitzer, K. and Karnosky, D. (1998) Experimental design and early performance of FACTS II (*Aspen FACE*) in Wisconsin, USA. *Abstracts of the GCTE-LUCC Open Science Conference on Global Change,* Barcelona, Spain, 14–18 March 1998, p. 57.

Janous, D., Dvorak, V., Oplustilova, M. and Kalina, J. (1996). Chamber effects and responses of trees in the experiment using open-top chambers. *Journal of Plant Physiology* **148**, 332–338.

Jones, P., Jones, J.W., Allen, L.H.J. and Mishoe, J.W. (1984). Dynamic computer control of closed environmental plant growth chambers. Design and verification. *Transactions of the ASAE* **27**, 879–888.

Jongen, M., Jones, M.B., Hebeisen, T., Blum, H. and Hendrey, G. (1995). The effects of elevated CO_2 concentrations on the root-growth of *Lolium perenne* and *Trifolium repens* grown in a FACE system. *Global Change Biology* **1**, 361–371.

Kellomaki, S. and Wang, K.Y. (1997). Effects of long-term CO_2 and temperature elevation on crown nitrogen distribution and daily photosynthetic performance of Scots pine. *Forest Ecology and Management* **99**, 309–326.

Kimball, B.A. (1983). Carbon dioxide and agricultural yield—an assemblage and analysis of 430 prior observations. *Agronomy Journal* **75**, 779–788.

Kimball, B.A. (1992). Cost comparisons among free-air CO_2 enrichment, open-top chamber, and sunlit controlled-environment chamber methods of CO_2 exposure. *Critical Reviews in Plant Sciences* **11**, 265–270.

Kimball, B.A., Lamorte, R.L., Peresta, G.J., Mauney, J.R., Lewin, K.F. and Hendrey, G.R. (1992). Weather, soils, cultural practices, and cotton growth data from the 1989 FACE experiment in IBSNAT format. *Critical Reviews in Plant Sciences* **11**, 271–308.

Kimball, B.A., Lamorte, R.L., Seay, R.S., Pinter, P.J., Rokey, R.R., Hunsaker, D.J., Dugas, W.A., Heuer, M.L., Mauney, J.R., Hendrey, G.R., Lewin, K.F. and Nagy, J. (1994). Effects of free-air CO_2 enrichment on energy-balance and evapotranspiration of cotton. *Agricultural and Forest Meteorology* **70**, 259–278.

Kimball, B.A., Pinter, P.J., Garcia, R.L., Lamorte, R.L., Wall, G.W., Hunsaker, D.J., Wechsung, G., Wechsung, F. and Kartschall, T. (1995). Productivity and water-use of wheat under free-air CO_2 enrichment. *Global Change Biology* **1**, 429–442.

Knapp, A.K., Cocke, M., Hamerlynck, E.P. and Owensby, C.E. (1994). Effect of elevated CO_2 on stomatal density and distribution in a C_4 grass and a C_3 forb under field conditions. *Annals of Botany* **74**, 595–599.

Koch, G.W. (1993). The use of natural situations of CO_2 enrichment in studies of vegetation responses to increasing atmospheric CO_2. In: *Design and Execution of Experiments on CO_2 Enrichment. Proceedings of a Workshop held at Weidenberg, Germany, 26–30 October 1992* (Ed. by E.D. Schulze and H.A. Mooney), pp. 381–391. Ecosystems Research Report 6, Commission of the European Communities/E. Guyot SA, Brussels.

Koch, K.E. (1996). Carbohydrate-modulated gene expression in plants. *Annual Review of Plant Physiology and Plant Molecular Biology* **47**, 509–540.

Leavitt, S.W., Paul, E.A., Kimball, B.A., Hendrey, G.R., Mauney, J.R., Rauschkolb, R., Rogers, H., Lewin, K.F., Nagy, J., Pinter, P.J. and Johnson, H.B. (1994). Carbon isotope dynamics of free-air CO_2-enriched cotton and soils. *Agricultural and Forest Meteorology* **70**, 87–101.

Lee, H.S.J. and Jarvis, P.G. (1996). The effects of tree maturity on some responses to elevated CO_2 in Sitka spruce (*Picea sitchensis* Bong. Carr.). In: *Carbon Dioxide and Terrestrial Ecosystems* (Ed. by G.W. Koch and H.A. Mooney), pp. 53–70. Academic Press, San Diego.

Lewin, K.F., Hendrey, G.R., Nagy, J. and Lamorte, R.L. (1994). Design and application of a free-air carbon dioxide enrichment facility. *Agricultural and Forest Meteorology* **70**, 15–29.

Lewis, C.E. (1994). *The Effects on Photosynthetic CO_2 Assimilation of Long-term Elevation of Atmospheric* CO_2 *Concentration: An Assessment of the Response of* Trifolium repens *L. cv.* Milkanova *in FACE.* M.Sc. thesis, University of Essex, Colchester.

Long, S.P. and Drake, B.G. (1991). Effect of the long-term elevation of CO_2 concentration in the field on the quantum yield of photosynthesis of the C_3 sedge *Scirpus olneyi*. *Plant Physiology* **96**, 221–226.

Long, S.P., Farage, P.K., Nie, G.Y. and Osborne, C.P. (1995). Photosynthesis and rising CO_2 concentration. In: *Photosynthesis: From Light to Biosphere,* Vol. 5 (Ed. by P. Mathis), pp. 729–736. Kluwer Academic Publishers, Dordrecht, The Netherlands.

Long, S.P., Nie, G.Y., Baker, N.R., Drake, B.G., Farage, P.K., Hendrey, G. and Lewin, K.H. (1992). The implications of concurrent increases in temperature, CO_2 and tropospheric O_3 for terrestrial C_3 photosynthesis. *Photosynthesis Research* **34**, 108–108.

Luscher, A., Hendrey, G.R. and Nosberger, J. (1998). Long-term responsiveness to free-air CO_2 enrichment of functional types, species and genotypes of plants from fertile permanent grassland. *Oecologia* **113**, 37–45.

McLeod, A.R. (1993). Open-air exposure systems for air pollutants studies—their potential and limitations. In: *Design and Execution of Experiments on CO_2 Enrichment. Proceedings of a Workshop held at Weidenberg, Germany, 26–30 October 1992* (Ed. by E.D. Schulze and H.A. Mooney), pp. 353–365. Ecosystems Research Report 6, Commission of the European Communities/E. Guyot SA, Brussels.

Masle, J., Farquar, G.D. and Gifford, R.M. (1990). Growth and carbon economy of wheat seedings as affected by soil resistance to penetration and ambient partial pressure of CO_2. *Australian Journal of Plant Physiology* **17**, 465–487.

Mauney, J.R., Lewin, K.F., Hendrey, G.R. and Kimball, B.A. (1992). Growth and yield of cotton exposed to free-air CO_2 enrichment (FACE). *Critical Reviews in Plant Sciences* **11**, 213–222.

Mauney, J.R., Kimball, B.A., Pinter, P.J., Lamorte, R.L., Lewin, K.F., Nagy, J. and Hendrey, G.R. (1994). Growth and yield of cotton in response to a free-air carbon dioxide enrichment (FACE) environment. *Agricultural and Forest Meteorology* **70**, 49–67.

Melillo, J., Callaghan, T.V., Woodward, F.I., Salati, E. and Sinha, S.K. (1990). Effects on ecosystems. In: *Climate Change: The IPCC Scientific Assessment* (Ed. by J.T. Houghton, G.J. Jenkins and J.J. Ephraums), pp. 283–310. Cambridge University Press, Cambridge.

Meyers, D.A., Thomas, R.B. and DeLucia, E.H. (1998). Photosynthetic capacity of loblolly pine (*Pinus taeda* L.) trees during the first year of carbon dioxide enrichment in a forest ecosystem. *Plant, Cell and Environment* **21** (submitted).

Miglietta, F., Raschi, A., Bettarini, I., Resti, R. and Selvi, F. (1993). Natural CO_2 springs in Italy a resource for examining long term response of vegetation to rising atmospheric CO_2 concentrations. *Plant Cell Environ.* **16**, 873–878.

Miglietta, F., Giuntoli, A. and Bindi, M. (1996). The effect of free-air carbon dioxide enrichment (FACE) and soil-nitrogen availability on the photosynthetic capacity of wheat. *Photosynthesis Research* **47**, 281–290.

Miglietta, F., Lanini, M., Bindi, M. and Magliulo, V. (1997). Free-air CO_2 enrichment of potato (*Solanum tuberosum* L.): Design and performance of the CO_2-fumigation system. *Global Change Biology* **3**, 417–427.

Monteith, J.L. (1978). A reassessment of maximum growth rates for C_3 and C_4 crops. *Experimental Agriculture* **14**, 1–5.

Murray, M.B., Leith, I.D. and Jarvis, P.G. (1996). The effect of long term CO_2 enrichment on the growth, biomass partitioning and mineral nutrition of Sitka spruce (*Picea sitchensis* (Bong) Carr). *Trees Structure and Function* **10**, 393–402.

Nagy, J., Lewin, K.F., Hendrey, G.R., Lipfert, F.W. and Daum, M.L. (1992). FACE facility engineering performance in 1989. *Critical Reviews in Plant Sciences* **11**, 165–185.

Nagy, J., Lewin, K.F., Hendrey, G.R., Hassinger, E. and LaMorte, R. (1994). FACE facility CO_2 concentration control and CO_2 use in 1990 and 1991. *Agricultural and Forest Meteorology* **70**, 31–48.

Nagy, J., Blum, H., Hendrey, G.R., Koller, S.R. and Lewin, K.F. (1995). *Reliability, CO_2 Concentration Control, and CO_2 Gas Use of the FACE Facility at ETH in 1993 and 1994.* Report No. BNL-61363. Biosystems and Process Sciences Division, Department of Applied Science, Brookhaven National Laboratory, Upton, NY.

Nakayama, F.S., Huluka, G., Kimball, B.A., Lewin, K.F., Nagy, J. and Hendrey, G.R. (1994). Soil carbon dioxide fluxes in natural and CO_2-enriched systems. *Agricultural and Forest Meteorology* **70**, 131–140.

Nie, G.Y. and Long, S.P. (1992). Effects of prolonged growth in elevated CO_2 concentrations in the field on the amounts of different leaf proteins. In: *Proceedings of IXth International Congress of Photosynthesis Research* (Ed. by N. Murata), pp. 855–858. Kluwer Academic, Dordrecht.

Nie, G.Y., Hendrix, D.L., Webber, A.N., Kimball, B.A. and Long, S.P. (1995a). Increased accumulation of carbohydrates and decreased photosynthetic gene transcript levels in wheat grown at an elevated CO_2 concentration in the field. *Plant Physiology* **108**, 975–983.

Nie, G.Y., Long, S.P., Garcia, R.L., Kimball, B.A., Lamorte, R.L., Pinter, P.J., Wall, G.W. and Webber, A.N. (1995b). Effects of free-air CO_2 enrichment on the development of the photosynthetic apparatus in wheat, as indicated by changes in leaf proteins. *Plant, Cell and Environment* **18**, 855–864.

Nijs, I., Ferris, R., Blum, H., Hendrey, G. and Impens, I. (1997). Stomatal regulation in a changing climate: A field study using Free Air Temperature Increase (FATI) and Free Air CO_2 Enrichment (FACE). *Plant, Cell and Environment* **20**, 1041–1050.

Nitschelm, J.J., Luscher, A., Hartwig, U.A. and VanKessel, C. (1997). Using stable isotopes to determine soil carbon input differences under ambient and elevated atmospheric CO_2. *Global Change Biology* **3**, 411–416.

Oechel, W.C., Cowles, S., Grulke, N., Hastings, S.J., Lawrence, B., Prudhomme, T., Riechers, G., Strain, B., Tissue, D. and Vourlitis, G. (1994). Transient nature of CO_2 fertilization in Arctic tundra. *Nature* **371**, 500–503.

Osborne, C.P., Long, S.P., Garcia, R.L., Wall, G.W., Kimball, B.A., Pinter, P.J., Lamorte, R.L. and Hendrey, G.R. (1995). Do shade and elevated CO_2 concentration have an interactive effect on photosynthesis—an analysis using wheat grown under Free-Air CO_2 Enrichment (FACE). In: *Photosynthesis: From Light to Biosphere* (Ed. by P. Mathis), Vol. 5, pp. 929–932. Kluwer Academic Publishers, Dordrecht, The Netherlands.

Osborne, C.P., LaRoche, J., Garcia, R.L., Kimball, B.A., Wall, G.W., Pinter, P.J., Lamorte, R.L., Hendrey, G.R. and Long, S.P. (1998). Does leaf position within a canopy affect acclimation of photosynthesis to elevated CO_2? Analysis of a wheat crop under Free-Air CO_2 Enrichment. *Plant Physiology* **117**, 1037–1045.

Owensby, C.E., Ham, J.M., Knapp, A.K., Bremer, D. and Auen, L.M. (1997). Water vapour fluxes and their impact under elevated CO_2 in a C_4 tallgrass prairie. *Global Change Biology* **3**, 189–195.

Pearson, M., Davies, W.J. and Mansfield, T.A. (1995). Asymmetric responses of adaxial and abaxial stomata to elevated CO_2—impacts on the control of gas-exchange by leaves. *Plant, Cell and Environment* **18**, 837–843.

Pinter, P.J., Kimball, B.A., Mauney, J.R., Hendrey, G.R., Lewin, K.F. and Nagy, J. (1994). Effects of free-air carbon dioxide enrichment on PAR absorption and conversion efficiency by cotton. *Agricultural and Forest Meteorology* **70**, 209–230.

Pinter, P.J., Kimball, B.A., Garcia, R.L., Wall, G.W., Hunsaker, D.J. and LaMorte, R.L. (1996). Free-air CO_2 enrichment: Responses of cotton and wheat crops. In: *Carbon Dioxide and Terrestrial Ecosystems* (Ed. by G.W. Koch and H.A. Mooney), pp. 215–249. Academic Press, San Diego.

Pleijel, H., Wallin, G., Karlsson, P.E., Skarby, L. and Sellden, G. (1994). Ozone deposition to an oat crop (*Avena sativa* L) grown in open-top chambers and in the ambient air. *Atmospheric Environment* **28**, 1971–1979.

Prior, S.A., Rogers, H.H. and Runion, G.B. (1993). Effects of free-air CO_2 enrichment on cotton root morphology. *Plant Physiology* **102**, 173.

Prior, S.A., Rogers, H.H., Runion, G.B. and Hendrey, G.R. (1994a). Free-air CO_2 enrichment of cotton—vertical and lateral root distribution patterns. *Plant and Soil* **165**, 33–44.

Prior, S.A., Rogers, H.H., Runion, G.B. and Mauney, J.R. (1994b). Effects of free-air CO_2 enrichment on cotton root growth. *Agricultural and Forest Meteorology* **70**, 69–86.

Prior, S.A., Rogers, H.H., Runion, G.B., Kimball, B.A., Mauney, J.R., Lewin, K.F., Nagy, J. and Hendrey, G.R. (1995). Free-air carbon dioxide enrichment of cotton—root morphological characteristics. *Journal of Environmental Quality* **24**, 678–683.

Prior, S.A., Torbert, H.A., Runion, G.B., Rogers, H.H., Wood, C.W., Kimball, B.A., LaMorte, R.L., Pinter, P.J. and Wall, G.W. (1997). Free-air carbon dioxide enrichment of wheat: soil carbon and nitrogen dynamics. *Journal of Environmental Quality* **26**, 1161–1166.

Rafarel, C.R., Ashenden, T.W. and Roberts, T.M. (1995). An improved solardome system for exposing plants to elevated CO_2 and temperature. *New Phytologist* **131**, 481–490.

Read, J.J., Morgan, J.A., Chatterton, N.J. and Harrison, P.A. (1997). Gas exchange and carbohydrate and nitrogen concentrations in leaves of *Pascopyrum smithii* (C_3) and *Bonteloua gracilis* (C_4) at different carbon dioxide concentrations and temperatures. *Annals of Botany* **79**, 197–206.

Reece, C.F., Krupa, S.V., Jager, H.-J., Roberts, S.W., Hastings, S.J. and Oechel, W.C. (1995). Evaluating the effects of elevated levels of atmospheric trace gases on herbs and shrubs: a prototype dual array field exposure system. *Environmental Pollution* **90**, 25–31.

Roberts, M.J., Long, S.P., Tieszen, L.L. and Beadle, C.L. (1993). Measurement of plant biomass and net primary production of herbaceous vegetation. In: *Photosynthesis and Production in a Changing Environment: A Field and Laboratory Manual* (Ed. by D.O. Hall, J.M. O. Scurlock, H.R. Bolhàr-Nordenkampf, R.C. Leegood and S.P. Long), pp. 1–21. Chapman & Hall, London.

Rogers, A., Bryant, J.B., Raines, C.A., Long, S.P., Blum, H. and Frehner, M. (1995). Acclimation of photosynthesis to rising CO_2 concentration in the field. Is it determined by source/sink balance. In: *Photosynthesis: From Light to Biosphere* (Ed. by P. Mathis), Vol. 5, pp. 1001–1004.

Rogers, A., Fischer, B.U., Bryant, J., Frehner, M., Blum, H., Raines, C.A. and Long, S.P. (1998). Acclimation of photosynthesis to elevated CO_2 under low N nutrition is effected by the capacity for assimilate utilisation. Perennial ryegrass under free-air CO_2 enrichment (FACE). *Plant Physiology* (in press).

Rogers, H.H. and Dahlman, R.C. (1993). Crop responses to CO_2 enrichment. *Vegetatio* **104**, 117–131.

Rogers, H.H., Prior, S.A. and O'Neill, E.G. (1992). Cotton root and rhizosphere responses to free-air CO_2 enrichment. *Critical Reviews in Plant Sciences* **11**, 251–263.

Rosenzweig, C. and Parry, M.L. (1994). Potential impact of climate-change on world food-supply. *Nature* **367**, 133–138.

Runion, G.B., Curl, E.A., Rogers, H.H., Backman, P.A., Rodriguezkabana, R. and Helms, B.E. (1994). Effects of free-air CO_2 enrichment on microbial populations in the rhizosphere and phyllosphere of cotton. *Agricultural and Forest Meteorology* **70**, 117–130.

Russell, E.W. (1961). *Soil Conditions and Plant Growth.* 9th edn. Longmans, London.

Sage, R.F. (1994). Acclimation of photosynthesis to increasing atmospheric CO_2—the gas-exchange perspective. *Photosynthesis Research* **39**, 351–368.

Schortemeyer, M., Hartwig, U.A., Hendrey, G.R. and Sadowsky, M.J. (1996). Microbial community changes in the rhizospheres of white clover and perennial ryegrass exposed to free-air carbon dioxide enrichment (FACE). *Soil Biology and Biochemistry* **28**, 1717–1724.

Schulze, E.D. and Tenhunen, J.D. (1993). Planning a FACE experiment in a mixed forest stand. In: *Design and Execution of Experiments on CO_2 Enrichment. Proceedings of a Workshop held at Weidenberg, Germany, 26–30 October 1992* (Ed. by E.D. Schulze and H.A. Mooney), pp. 371–377. Ecosystems Research Report 6, Commission of the European Communities/E. Guyot SA, Brussels.

Sellers, P.J., Bounoua, L., Collatz, G.J., Randall, D.A., Dazlich, D.A., Los, S.O., Berry, J.A., Fung, I., Tucker, C.J., Field, C.B. and Jensen, T.G. (1996). Comparison of radiative and physiological-effects of doubled atmospheric CO_2 on climate. *Science* **271**, 1402–1406.

Sellers, P.J., Dickinson, R.E., Randall, D.A., Betts, A.K., Hall, F.G., Berry, J.A., Collatz, G.J., Denning, A.S., Mooney, H.A., Nobre, C.A., Sato, N., Field, C.B. and Henderson Sellers, A. (1997). Modeling the exchanges of energy, water, and carbon between continents and the atmosphere. *Science* **275**, 502–509.

Senock, R.S., Ham, J.M., Loughin, T.M., Kimball, B.A., Hunsaker, D.J., Pinter, P.J., Wall, G.W., Garcia, R.L. and Lamorte, R.L. (1996). Sap flow in wheat under free-air CO_2 enrichment. *Plant, Cell and Environment* **19**, 147–158.

Sheen, J. (1994). Feedback-control of gene-expression. *Photosynthesis Research* **39**, 427–438.

Spring, G.M., Priestman, G.H. and Grime, J.P. (1996). A new field technique for elevating carbon dioxide levels in climate change experiments. *Functional Ecology* **10**, 541–545.

Stirling, C.M., Davey, P.A., Williams, T.G. and Long, S.P. (1997). Acclimation of photosynthesis to elevated CO_2 and temperature in five British native species of contrasting functional type. *Global Change Biology* **3**, 237–246.

Stitt, M. (1991). Rising CO_2 levels and their potential significance for carbon flow in photosynthetic cells. *Plant, Cell and Environment* **14**, 741–762.

Tissue, D.T., Thomas, R.B. and Strain, B.R. (1996). Growth and photosynthesis of Loblolly pine (*Pinus taeda*) after exposure to elevated CO_2 for 19 months in the field. *Tree Physiology* **16**, 49–59.

Unsworth, M.H., Heagle, A.S. and Heck, H.H. (1984). Gas exchange in open-top field chambers—I. Measurement and analysis of atmospheric resistances to gas exchange. *Atmospheric Environment* **18**, 373–380.

Vu, J.C.V., Allen, L.H., Boote, K.J. and Bowes, G. (1997). Effects of elevated CO_2 and temperature on photosynthesis and Rubisco in rice and soybean. *Plant, Cell and Environment* **20**, 68–76.

Walklate, P.J., Xu, Z.G. and McLeod, A.R. (1996). A new gas injection method to enhance spatial utilization within a Free-Air CO_2 Enrichment (FACE) system. *Global Change Biology* **2**, 75–78.

Webber, A.N., Nie, G.Y. and Long, S.P. (1994). Acclimation of photosynthetic proteins to rising atmospheric CO_2. *Photosynthesis Research* **39**, 413–425.
Wechsung, G., Wechsung, F., Wall, G.W., Adamsen, F.J., Kimball, B.A., Garcia, R.L., Pinter, P.J. and Kartschall, T. (1995). Biomass and growth-rate of a spring wheat root-system grown in free-air CO_2 enrichment (FACE) and ample soil-moisture. *Journal of Biogeography* **22**, 623–634.
Whitehead, D., Hogan, K.P., Rogers, G.N.D., Byers, J.N., Hunt, J.E., McSeveny, T.M., Hollinger, D.Y., Dungan, R.J., Earl, W.B. and Bourke, M.P. (1995). Performance of large open-top chambers for long-term field investigations of tree response to elevated carbon dioxide concentration. *Journal of Biogeography* **22**, 307–313.
Wood, C.W., Torbert, H.A., Rogers, H.H., Runion, G.B. and Prior, S.A. (1994). Free-air CO_2 enrichment effects on soil carbon and nitrogen. *Agricultural and Forest Meteorology* **70**, 103–116.
Woodrow, I.E. (1994a). Control of steady-state photosynthesis in Sunflowers growing in enhanced CO_2. *Plant, Cell and Environment* **17**, 277–286.
Woodrow, I.E. (1994b). Optimal acclimation of the C_3 photosynthetic system under enhanced CO_2. *Photosynthesis Research* **39**, 401–412.
Xu, Z.G., Walklate, P.J. and McLeod, A.R. (1997). Numerical study of a full-size free-air fumigation system. *Agricultural and Forest Meteorology* **85**, 159–170.
Zanetti, S. and Hartwig, U.A. (1997). Symbiotic N_2 fixation increases under elevated atmospheric pCO_2 in the field. *Acta Oecologica—International Journal of Ecology* **18**, 285–290.
Zanetti, S., Hartwig, U.A., Luscher, A., Hebeisen, T., Frehner, M., Fischer, B.U., Hendrey, G.R., Blum, H. and Nosberger, J. (1996). Stimulation of symbiotic N_2 fixation in *Trifolium repens* L under elevated atmospheric pCO_2 In a Grassland Ecosystem. *Plant Physiology* **112**, 575–583.
Ziska, L.H. and Bunce, J.A. (1997). The role of temperature in determining the stimulation of CO_2 assimilation at elevated carbon dioxide concentration in soybean seedlings. *Physiologia Plantarum* **100**, 126–132.

APPENDIX

Abbreviation for FACE Facilities and Operators

BNL	Brookhaven National Laboratory
ELM	enhanced local mixing
ETH	Eidgenössische Technische Hochschule
ETHZ	Eidgenössische Technische Hochschule, Zürich
FACE	free-air carbon dioxide enrichment
FACTS	forest-atmosphere carbon transfer and storage
FFP	Forest FACE prototype
GCTE	global change and terrestrial ecosystems
IGBP	International Geosphere-Biosphere Programme
MAC	Maricopa Agricultural Center
NDFF	Nevada Desert FACE Facility
NTS	Nevada Test Site (of the NDFF)
OTC	open-top chamber
SPAR	soil, plant, atmosphere research chamber

Symbols and Other Terms

ET	potential evapotranspiration
ET_c	cumulative ET
ϵ_c	efficiency of conversion of intercepted light into biomass
ϵ_i	efficiency of interception of incident light by a plant canopy
F	CO_2 use rate (kg min^{-1})
h	solar elevation (degrees)
LAI	leaf area index
LHC	light-harvesting protein complex
pCO$_2$	partial pressure of carbon dioxide (Pa)
PID	proportional integral differential
Rubisco	ribulose bisphosphate carboxylase oxygenase
U	wind velocity (m s^{-1})
$V_{c,max}$	maximum velocity of carboxylation of Rubisco within a leaf (μmol m^{-2} s^{-1})

Modelling Terrestrial Carbon Exchange and Storage: Evidence and Implications of Functional Convergence in Light-use Efficiency

S.J. GOETZ AND S.D. PRINCE

I. SUMMARY

The practicality of ecological research at a global scale is increasing as a result of the development of satellite remote sensing which enables surface conditions to be inferred for large areas at high spatial and temporal resolutions. Remote sensing of canopy light absorption, for example, enables an important component of canopy carbon assimilation rates to be estimated over extensive areas. Modelling of carbon exchange and net storage may also be facilitated, owing to natural selection for a narrow range of light-use effi-

ADVANCES IN ECOLOGICAL RESEARCH VOL. 28
ISBN 0–12–013928–6

ciencies among a wide range of plant functional types. This "functional convergence" arises from resource allocation strategies that appear to maximise the benefits of carbon (energy) and nutrients (e.g. mainly nitrogen) relative to the costs of acquisition. Convergence is most evident on a leaf mass per unit ground area basis, a measure that more closely reflects the costs of resource acquisition than does leaf area. We demonstrate, however, that light absorption and utilisation are decoupled so that convergence is to be expected on gross production rather than net production, owing to differences in respiratory costs associated with synthesis and maintenance of plant constituents and associated "payback intervals" on carbon investment in different functional types. A link between fitness and carbon gain is noted in relation to gross primary production. We conclude that, while functional convergence provides a basis for the use of remote sensing of light absorption in measurement of primary production, models driven with light absorption need to include terms that describe the actual respiratory costs of maintenance and synthesis. Quantification of these processes will improve large area primary production models, and enhance the value of the information that can be acquired by remote sensing.

II. INTRODUCTION

Net primary production (NPP) is an important component of the global carbon cycle because it provides a measure of the amount of CO_2 removed from the atmosphere through net carbon exchange (photosynthesis less respiration). NPP is difficult to measure over large areas owing to spatial variability of environmental conditions and limitations in the accuracy of allometric equations used to estimate carbon gain from tree dimensions and other direct measurement techniques. If NPP could be mapped more accurately over large regions, significant uncertainties in the global carbon cycle (see IPCC, 1996; Keeling *et al.*, 1996) and discrepancies in the location and magnitude of terrestrial carbon sinks (Ciais *et al.*, 1995; Denning *et al.*, 1995; Lloyd and Farquhar, 1996) are likely to be resolved. Maps of NPP over large areas are also of relevance to issues of land-use change, food security and vegetation feedback on climate (Prince *et al.*, 1994). Towards these ends, various spatially explicit carbon-exchange models have been developed that have large spatial coverage. These can be broadly classified into statistical correlation models (Bazilevich *et al.*, 1971; Leith and Box, 1977; Esser, 1991), mechanistic physiological models (Melillo *et al.*, 1993; Warnant *et al.*, 1994; Foley *et al.*, 1996), and top-down or diagnostic models (Tans *et al.*, 1990; Ciais *et al.*, 1995; Keeling *et al.*, 1995). All of these models require spatially explicit input data, typically including meterological observations, soil maps, vegetation maps or remotely sensed vegetation observations.

Remotely sensed data are spatially contiguous and are made no more than a few days apart, unlike the interpolations from widely separated point observations and static maps that are used to provide input variables for many carbon-exchange models. Spectral reflectance properties of vegetation measured with satellite remote sensing have been related to NPP in a number of different biomes at a variety of spatial scales, from the plot level (Asrar *et al.*, 1985; Daughtry *et al.*, 1992; Goetz and Prince, 1996) to the globe (Potter *et al.*, 1993; Prince and Goward, 1995). The physical basis for the observed correlations between spectral reflectance and NPP is the existence of a relationship between reflection and absorption of solar radiation by vegetation canopies, and between the amount of absorbed photosynthetically active radiation (APAR) and its utilisation for NPP. Both of these processes (light absorption and utilisation) vary in time and space, however, and must be adequately modelled in order to use remote sensing to estimate NPP.

The measurement of APAR using remotely sensed reflectance in the red and near infrared has been the subject of a great deal of research in the past decade (e.g. Kumar and Monteith, 1982; Sellers *et al.*, 1992; Goel and Qin, 1994). The ratio of NPP to APAR (often called the production "efficiency", and given the symbol ϵ; units g C MJ^{-1}) has become an important parameter because of its implications for the use of remote sensing to estimate both net and gross primary production (GPP) (Russel *et al.*, 1989; Prince, 1991a; Landsberg *et al.*, 1997). If the values of ϵ were similar for all plant types and biomes, then the task of modelling net carbon exchange with remote sensing would be simple because NPP would be directly proportional to light harvesting. If, however, ϵ varied widely, then representative values would have to be determined for each vegetation type or biome on a case by case basis (e.g. Ruimy *et al.*, 1994). Assigning values of ϵ on the basis of biome type assumes between-biome variability in ϵ is greater than within-biome variability, which may not be realistic (e.g. Goetz and Prince, 1996; Landsberg *et al.*, 1997). Alternatively ecophysiological simulation models, requiring explicit parameterisations, for which a number of non-trivial assumptions are required, may be used in conjunction with the remote sensing observations to derive values of ϵ (Running and Hunt, 1993; Goetz and Prince, 1998).

This paper explores simplifying assumptions about plant physiology that can be justified on the basis of evolutionary theory in order to exploit the capabilities of satellite remote sensing data to drive models of regional to global scale C uptake in NPP. It has been suggested that limited resource availability in the natural environment and high resource acquisition costs result in an optimisation of resource allocation by plants through natural selection, which results in maximisation of carbon gain and convergence on a narrow range of ϵ (Field, 1991). If true, this "functional convergence" hypothesis would provide strong theoretical support for an interpretation of the observed correlation between remotely sensed APAR and NPP in terms of evolutionary optimisation. We explore the evidence for such convergence and the implications for global estimation of carbon exchange and storage using remote sensing. We first review the tenets of the functional convergence hypothesis, noting cases where it may fail for a number of different reasons and establishing the criteria by which it may be assessed.

III. THE FUNCTIONAL CONVERGENCE HYPOTHESIS AND BASES FOR ITS ASSESSMENT

The functional convergence hypothesis (Field, 1991) states that plant canopy light harvesting is scaled to the availability of all other resources such that "biochemical capacity for CO_2 fixation is curtailed whenever a limitation in the availability of any resource prevents the efficient exploitation of addition-

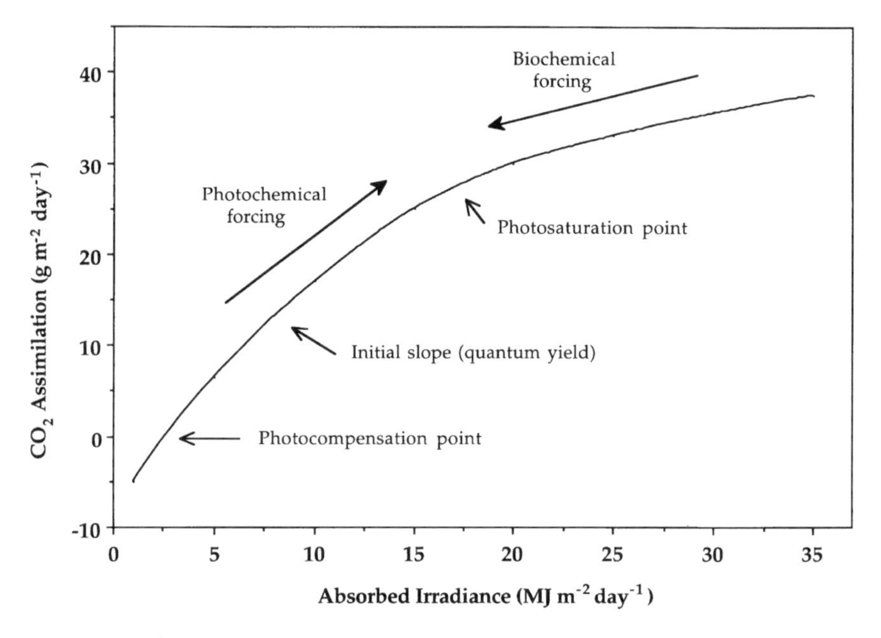

Fig. 1. An idealised photosynthesis–light response curve indicating the key features of net photosynthetic rates relative to absorbed photosynthetically active radiation. Photocompensation point delimits the region above which net CO_2 exchange is positive for the leaf (or plant canopy). Quantum yield is defined by the initial (typically linear) slope of the light response curve along which photosynthesis is limited by light reactions. Photosaturation point defines the region above which photosynthesis is limited by biochemical capacity (i.e. carboxylation rates). Functional convergence suggests plant canopies have been tuned by natural selection (indicated by arrows) to operate, over the longer term (months to years), near the photosaturation point, effectively balancing resource limitations (light harvesting versus biochemical capacity).

al capacity." In other words, despite light saturation of individual leaves (i.e. beyond their CO_2 compensation point in a photosynthesis–light response curve; Figure 1), plant canopies tend not to have excess photosynthetic capacity because selection favours investment in acquisition of the most limiting resource and not in additional photosynthetic machinery (i.e. leaves and chloroplasts) (Field, 1983; Sellers *et al*, 1992; Haxeltine and Prentice, 1996).

Functional convergence implies that the scaling of light harvesting to resource availability among plant functional types is independent of the nature of the limiting resource. Field (1991) notes that "only if investments in light harvesting reflect the annual integral of the availability of other resources should the annual integral of APAR predict growth accurately." Thus, evidence that various functional types have similar light-use efficiency (i.e. light harvested : production ratio) is taken to suggest that the type of stress is inconsequential. In this way Field views biochemical capacity for CO_2 fixation as "a master integrator of the environment".

Functional convergence may fail to occur under some conditions. For example, interannual variability in resource availability may limit the ability of plants to scale the capacity for resource uptake in any given year and this may vary between functional types (see Goward and Prince, 1995). Similarly, short-term variations in photosynthesis may not scale with light harvesting or moisture availability. Most obviously, evergreen species in seasonal habitats (e.g. non-deciduous boreal conifers) would be expected to diverge most from functional convergence, owing to long periods of the year when light is absorbed but the plants are photosynthetically dormant.

In order to carry out a direct test of the functional convergence hypothesis we need to measure canopy PAR absorption and primary production. Since functional convergence is based on evolutionary arguments, the measurements may have to be prolonged over a full growing season and, ideally, the entire life histories of the components of a community. PAR absorption can be measured over long periods with arrays of sensors above and below the canopy (e.g. Bégué *et al.*, 1996), although absorption by non-photosynthetic components may have to be estimated by techniques such as artificial defoliation. The measurement of primary production, on the other hand, is subject to great difficulties. Harvest methods often omit significant components such as below ground carbon allocation and fixed carbon lost in exudates and volatile carbon compounds (Prince *et al.*, 1994). CO_2 fluxes measured by chamber enclosure methods also miss these components. Both chamber and harvest methods have problems in scaling up to entire communities. For reasons discussed below, we require gross (GPP), not net primary production (NPP) and so additional measurements are needed to estimate autotrophic respiration in all living parts of the community. Eddy correlation CO_2 flux methods estimate net ecosystem production (NEP), which includes autotrophic and heterotrophic respiration and, without additional measurements that involve significant uncertainties, we are unable to use such data to estimate NPP.

Not only are measurements of APAR and production needed for an adequate test of the hypothesis, but also ratios of these two variables are needed under conditions in which there are variations in resource availability. Such data do not exist, although more comprehensive programmes of carbon balance studies are being proposed. Nevertheless, the implications of the functional convergence hypothesis for regional monitoring of primary production using remotely sensed observations are so important that it is worthwhile exploring the circumstantial evidence that is available to build confidence in the assumption of its validity until empirical evidence can be assembled. Since evolutionary optimisation is proposed as the driving mechanism of functional convergence, the appropriate timescale is lifetime, not hours or even years in the case of a perennial. Nevertheless, light is a fleeting resource that must be captured when available or it is lost, and so optimal light harvesting may be expected throughout the life of a plant. The shortest appropriate time interval is set by the time taken to adjust the capacity for light absorption, days in the case of the pigment system and weeks in the case of the structure of the leaf canopy. At shorter time intervals deviations from the correlation of light absorption and primary production may be expected as the scaling of light absorption is unable to track more rapid fluctuations in resources. Also evolution acts on the breeding population (single species), whereas most canopies are composed of more than one plant. In using remote sensing to estimate NPP, we are addressing the community scale, not that of individual plants. Thus two distinct issues are involved, firstly, the response of individual plants to their environment that we suppose to be determined by evolutionary processes and, secondly, those processes that determine community composition which leads to broad exploitation of resources by different species having complementary growth forms and life histories, so long as any species can function in the environment.

In order to supplement the limited evidence available from field measurements and simulations of ϵ we have assessed the evidence for convergence in ϵ among functional types from evolutionary ecology, particularly those aspects concerned with resource allocation strategies and resource use efficiencies, and the relationship of these to carbon gain among functional types. We use these lines of evidence to consider the association between carbon cost–benefit ratios and the payback interval on carbon investment, and the manner in which these vary among plant functional types.

We have approached such an assessment by evaluating the evidence available from a variety of sources, including plant physiology, evolutionary biology and environmental physics. In particular, we consider the extent to which light-use efficiency can be regarded as the result of efficient resource use and allocation, involving trade-offs, in the context of limited resources. We specifically evaluate the hypothesis in relation to remotely sensed canopy light absorption by establishing the links between various suites of traits that are associated with different functional types. We then consider the trade-offs that result in different cost–benefit ratios and "payback intervals" and explore the evidence for co-ordinated use of multiple limiting resources. Finally, we examine how resource-use co-ordination can lead to maximised carbon gain, how carbon gain relates to fitness, and what the implications of the various findings are for modelling terrestrial primary production with remotely sensed observations. We begin with a brief summary of the links between remotely sensed canopy reflectance, light absorption and net primary production and establish a range of observed values of light-use efficiency in terms of *net* production.

IV. LINKS BETWEEN REMOTE SENSING, LIGHT ABSORPTION AND PRIMARY PRODUCTION

Spectral vegetation indices (SVI) are derived from spectral reflectances of vegetation measured by radiometers above the canopy on various platforms, typically aircraft or Earth-orbiting satellites. An SVI image of Africa obtained from the Advanced Very High Resolution Radiometer (AVHRR) on the NOAA meteorological satellites (Figure 2) shows high values associated with dense humid forest and low values associated with sparse vegetation cover, for example, through the Sahelian zone and into the Sahara. The relationship of SVIs to vegetation cover has been used to estimate vegetation properties, such as leaf area index (LAI) (Curran, 1983; Running and Nemani, 1987; Goetz, 1997) and the fraction of incident photosynthetically active radiation (F_{par}) intercepted or absorbed by vegetation canopies (Asrar *et al.*, 1984; Sellers, 1985,1987; Goetz and Prince, 1996). Sensitivity studies of these relationships have demonstrated various degrees of non-linearity (Choudhury, 1987; Goward and Huemmrich, 1992; Goel and Qin, 1994), which, besides affecting the scale dependence of the relationship, demonstrate a sensitivity to different vegetation cover types. F_{par} has, in turn, been shown to be robustly linked to photosynthetic potential (ρ_c, the maximum rate of CO_2 assimilation per unit leaf area under ambient CO_2 concentration) and the slope of unstressed stomatal conductance relative to incident PAR (∇_f), both from theoretical considerations and in field measurements (Verma *et al.*, 1993; Sellers *et al.*, 1995).

As a result of these statistical relationships, annual sums of the product of SVIs and incident PAR (whether measured at the surface with pyranometers or modelled with remotely sensed observations, e.g. Dye and Shibasaki (1995)) have been positively correlated with annual NPP in crops (Monteith, 1977; Tucker *et al.*, 1981; Steven *et al.*, 1983), semi-arid grasslands (Tucker *et al.*, 1983; Prince and Astle, 1986; Prince, 1991b), forest stands (Runyon *et al.*,

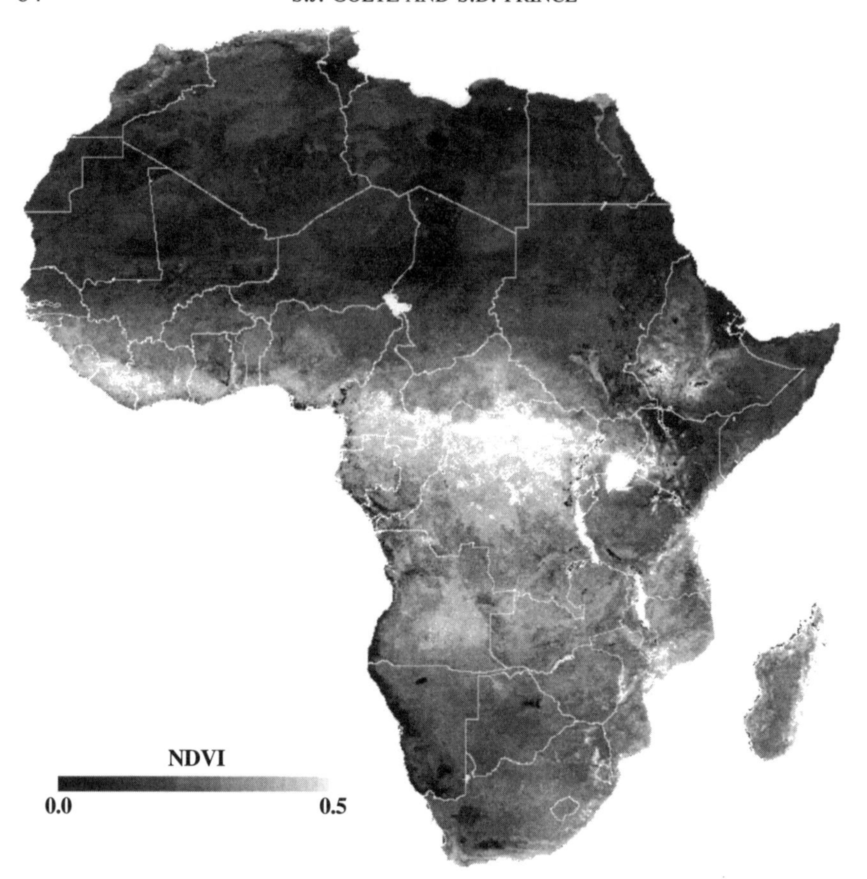

Fig. 2. A composited spectral vegetation index (SVI) image of Africa derived from 72 monthly maximum value images (for the time period 1982–1987). The SVI shown is the commonly used "normalised difference" vegetation index (NDVI), defined as the difference between infrared and visible reflectance divided by the sum of the two. Unpublished image provided by Nadine Laporte.

1994; Goetz and Prince, 1996), and even at continental to global scales (Goward *et al.*, 1985; Potter *et al.*, 1993; Prince and Goward, 1995). In addition, seasonal oscillations in continental-scale atmospheric CO_2 concentration have been correlated with the dynamics of spectral vegetation indices (Tucker *et al.*, 1986), as well as with semi-empirical models that assume constant ϵ (Hunt *et al.*, 1996; Knorr and Heimann, 1996). While these studies demonstrate a link between spectral vegetation indices and carbon-exchange dynamics, there are significant uncertainties in the estimates of NPP made from light absorption alone, and therefore in the range of ϵ (Prince, 1991a; Goetz and Prince, 1996; Landsberg *et al.*, 1997).

The theoretical range of annual ϵ is between 0.5 and 4.0 g C MJ^{-1}, as calculated from the quantum yield (ϕ) (Figure 1), the CO_2 photocompensation point, internal leaf CO_2 concentration, and the CO_2/O_2 specificity ratio (Collatz *et al.*, 1991; Prince and Goward, 1995). Empirically derived values of ϵ range from about 0.2 to 1.8 g C MJ^{-1} among a range of biomes, crop varieties and forest stands (summarised in Figure 3). Simulated values derived from an ecophysiological model of plant growth, driven by meteorological data, have a similar range of values (0.2–1.95 g C MJ^{-1}) for a variety of ecosystems in North America (Running and Hunt, 1993). Results from global primary production models (Potter *et al.*, 1993; Ruimy *et al.*, 1994; Prince and Goward, 1995) occupy a somewhat narrower range (0.1–1.0 g C MJ^{-1}), probably as a result of spatial averaging. The observed range in ϵ is known to be affected by different measures of radiation (incident, intercepted or absorbed) and production (above-ground, below-ground or total NPP), as well as photosynthetic pathway (C3, C4) thus comparisons must account for these differences (Prince, 1991a). Other factors, including measurement error, are also involved. As a result of observed variability in ϵ there has been recent debate on the utility of ϵ in NPP and crop yield models (see Monteith, 1994). The variability in ϵ and inconsistencies in its measurement also suggest insufficient evidence to either support or detract from the concept of functional convergence.

Several studies have shown, however, that the range in annual ϵ can be greatly reduced by accounting for environmental physiology, respiration and other carbon losses (Runyon *et al.*, 1994; Landsberg *et al.*, 1997; Goetz and Prince, 1998). Accounting for stress in measured or derived values of ϵ has been accomplished by a variety of methods, the simplest of which involve modelling variables that affect stomatal control (e.g. air temperature, vapour pressure deficit, soil moisture) as scalars that reduce potential photosynthesis (Landsberg, 1986; Prince, 1991a; Goward *et al.*, 1994; Prince and Goward, 1995). There has been little work to account for variations in ϵ introduced by herbivory, disease or differences in respiratory costs.

Ambiguities in light-use efficiency terminology led Prince (1991a) to make a distinction between ϵ measured on the basis of net production (ϵ_n) or gross production (ϵ_g) and between environmentally stressed (ϵ) or hypothetical unstressed or maximum (ϵ^*) values of these. A narrow range in ϵ, after accounting for environmental factors, suggests it may not be necessary to establish unique values of ϵ for different vegetation types in order to estimate NPP with remotely sensed measurements. It has been difficult to confirm this possibility, however, owing to a paucity of comparable measured values. For example, the values of ϵ_n reported in the literature are rarely whole ecosystem values, and do not include either below-ground production and under-storey or ground-cover vegetation, nor do they allow for the effects of herbivory, decomposition and other carbon losses. We note that recent efforts under the auspices of the International Geosphere-Biosphere Programme

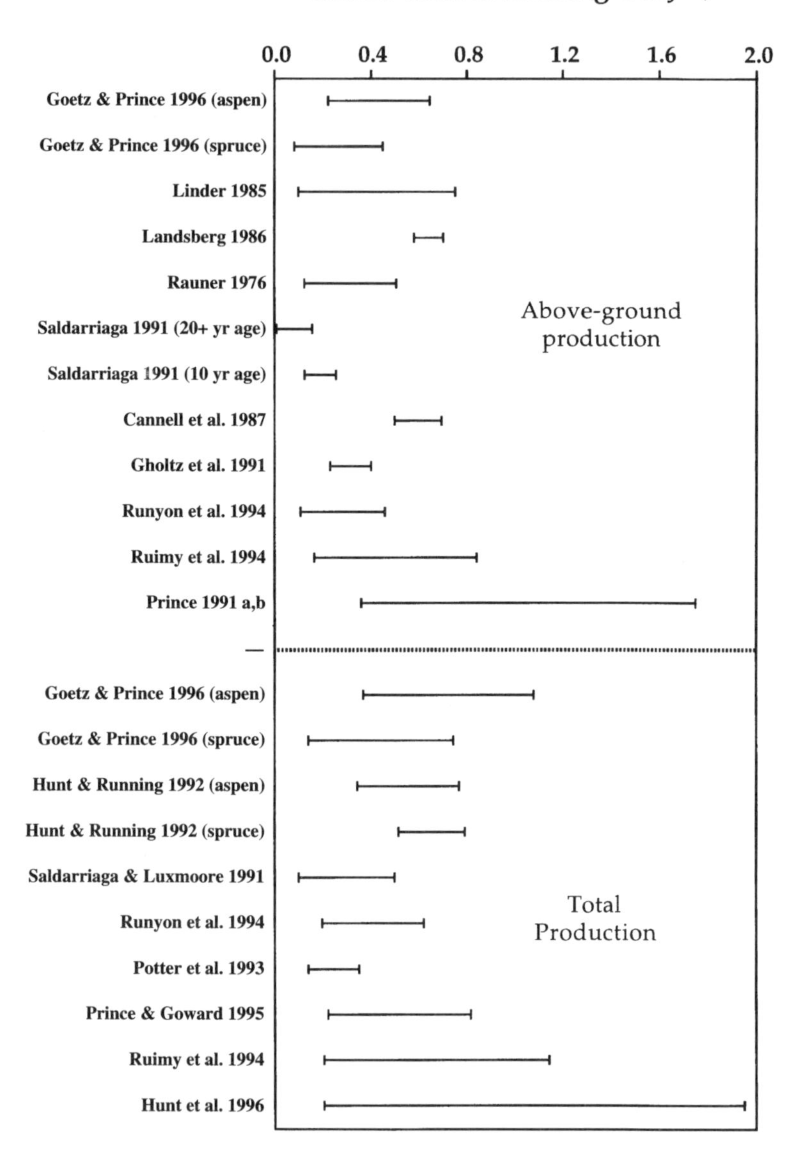

Fig. 3. A comparison of the reported range in ϵ the carbon yield of APAR for a number of studies, primarily forest stands. The radiation and production measures vary between studies which makes such comparisons difficult (Prince, 1991a). The values are categoried into total versus above-ground production. Additional information on these data are reported in Goetz and Prince (1996).

(IGBP) Global Primary Production Data Initiative (GPPDI) have focused on addressing these disparities in NPP data sets (Prince *et al.*, 1994).

V. RESOURCE CONSTRAINTS AND ADAPTIVE STRATEGIES

Adaptive strategies, defined as suites of co-adapted characteristics, arise as natural selection acts on a range of traits in a population. Strategies that are stable in an evolutionary sense, or have a "dynamic stability" are those most likely to persist (Maynard Smith, 1978). There is ample evidence that trade-offs in resource allocation result in specific and predictable suites of life-history traits (Gadgil and Bossert, 1970; Stearns, 1976, 1989; Ricklefs, 1991). These suites of traits have been classified in a number of ways, including MacArthur and Wilson's (1967) widely applied model of *r* versus *K* selection, Grime's (1979) triangle model of primary plant growth strategies, and Tilman's (1988) model of biomass partitioning and relative growth rate (RGR; the product of net assimilation rate, specific leaf area and leaf weight ratio, where specific leaf area is leaf area per unit leaf weight, and leaf weight ratio is the fraction of total plant biomass allocated to foliage).

There has been a debate on the validity and utility of these classifications of strategies (Shipley and Peters, 1990; Tilman, 1991; Silvertown *et al.*, 1992) but there is little doubt that many plants have suites of traits that are related to characteristics of their environment. For example, studies at the leaf, plant, stand and ecosystem level demonstrate that the lifespan of a plant's foliage is closely related to other traits in the suite, including ρ_c, whole-plant lifespan, and RGR (Lambers and Poorter, 1992; Reich *et al.*, 1992, 1997). The relation between ρ_c and RGR has long been noted in, and contrasted between, needleleaf evergreen and broadleaf deciduous trees grown under similar conditions (Table 1), and is documented in data compiled for 159 tree species of North America (Loehle, 1988). Other associations between life history traits have been reported (Stearns, 1989; Ricklefs, 1991; Chapin, 1993). Although individual traits of a given suite may not always be present, the association of traits is constant and is generally considered a result of their co-evolution (Partridge and Harvey, 1988) or their sequential evolution within a clade (i.e. a group of organisms derived from a common ancestor). Plant functional type classifications associated with various suites of traits have been proposed for a range of spatial scales from individual plants to the globe (Box, 1996; Smith *et al.*, 1997). Some classifications have been defined with remotely sensed observations (e.g. Defries *et al.*, 1995; Nemani and Running, 1997).

Owing to the linkages of traits in different plant functional types, several plant variables have been used as diagnostic characteristics or "ecological integrators". Vitousek (1982) and Field and Mooney (1986) view nitrogen

Table 1
Different properties and process rates of boreal tree species (*Picea mariana*, *Populus tremuloides*) representing contrasted functional types

Picea mariana Relative Rate or Allocation	Variable	*Populus tremuloides* Relative Rate or Allocation
Low	Photosynthetic Capacity	High
Low	Relative Growth Rate	High
Low	Foliage Nitrogen	High
Slow	Nutrient Turnover	Fast
High	Leaf Mass/Area	Low
High	Foliage Carbon : Nitrogen Ratio	Low
High	Respiration : Assimilation Ratio	Low
High	Non-growth Resource Allocation	Low
Low	Net Primary Production	High
Long	Lifespan	Short

utilisation as one such integrator. Reich *et al.* (1992) identify foliage longevity as an integrator, "one of several inter-related traits that cannot be viewed in isolation". Chapin (1993) regards RGR as the "central figure of plant adaptive strategies", which allows it to be used "as an index of other features". Notable exceptions to these generalisations may occur as plants age or change successional status (Lerdau, 1992; Gleeson and Tilman, 1994). For example, there is evidence that plant age and size influence linkages between traits (Shipley and Peters, 1990; Chapin, 1993). Thus the trait relatedness expressed in different growth-forms is determined by dynamic resource allocation strategies which may change with ontogeny or changes in the environment (Kawecki, 1993; Perrin and Sibley, 1993). This subject is revisited later in relation to trade-offs in resource-use efficiency.

We have presented evidence that: (1) resource limitations result in a number of different adaptive strategies that reflect trade-offs in resource allocation; (2) suites of traits can be broadly classified in plant functional types; and (3) these can be identified by any one of a number of integrative traits. We next consider the carbon costs and benefits associated with different functional types in order to better characterise the relation between (and the relevance of) gross versus net carbon gain among functional types.

VI. ALLOCATION, DEFENCE COSTS AND PAYBACK INTERVALS

Allocation of assimilated carbon to different plant components is an important consideration in relation to NPP because allocation to photosynthetic organs produces a positive feedback in additional carbon assimilation, whereas costs of assimilate allocation to non-photosynthetic components are not negligible and vary among functional types. In trees it has been suggested that carbon allocation follows a hierarchy, where the first priority is given to maintenance respiration of the living tissues, next to production of foliage and fine roots, then to flower and seed production (which may slow growth of other components), then to primary growth (terminal and lateral branch growth and root extension), and, finally, to addition of xylem and defensive compounds (Oliver and Larson, 1990). This resource partitioning hierarchy results in enhanced allocation to those plant components associated with extraction of limiting resources (or reproduction). One of the characteristics of fast-growing species is that they have a greater ability to adjust allocation, for example, in response to stress (Kachi and Rorison, 1989; van der Werf *et al.*, 1993; Laurence *et al.*, 1994). Such observed allocation patterns may be best viewed as "adaptively plastic responses to resource gradients" (Gleeson and Tilman, 1994), which is particularly evident in specific leaf area (Lambers and Poorter, 1992).

Differences in allocation and plasticity between functional types are linked to investment in secondary compounds required for defence, that is, for protection against herbivores and pathogens (Field and Mooney, 1986; Fagerström *et al.*, 1987; Loehle, 1988). Allocation to chemical defence compounds (including resins and phenolics such as lignin and tannins) can range from 10% to 30% of dry weight in herbaceous species and even more in woody species (e.g. Coley *et al.*, 1985). Chemical defences are of particular significance in resource utilisation because they require a higher allocation of assimilate and have higher respiration costs for synthesis relative to cellulose. The release of CO_2 associated with the synthesis of defensive chemicals, for example, is 9–15 times that associated with sucrose synthesis and 4–10 times that associated with cellulose synthesis (Penning de Vries, 1975). CO_2 release from the synthesis of volatile terpenoids that are associated with leaf protection in high-radiation regimes is even greater.

Economic analogies have frequently been invoked that describe these allocation patterns in terms of trade-offs between costs and returns (Bloom *et al.*, 1985; Chapin *et al.*, 1987; Lerdau *et al.*, 1994). Although the costs versus the benefits of defence processes are difficult to quantify (Raven, 1986), observed negative correlations between lifespan and RGR have been attributed to the high energetic cost of defensive chemical synthesis and the diversion of assimilated carbon from current growth (Chapin *et al.*, 1987; Lambers and Poorter, 1992; Lerdau, 1992). Loehle (1988) showed that the allocation to defence in relation to growth rate and longevity could account for observed differences in tree-growth rates through mechanisms by which species either defend themselves (slow-growing, long-lived) or outgrow pathogens (fast-growing, short-lived). Similar correlations have been noted at the leaf level (Chabot and Hicks, 1982; Coley, 1986; Harper, 1989).

Different relative growth rates and carbon allocation patterns between functional types may therefore be stated in economic terms in which carbon respiratory costs are balanced against assimilation gains. Variability in the ratio of annual carbon respiration to assimilation has been directly related to variability in the carbon yield of APAR (i.e. ϵ_n) of different boreal tree species (Figure 4; Goetz and Prince, 1998). This finding is consistent with Loehle's (1988) observation that increased respiration costs are associated with differences in allocation strategies between functional types, which we refer to as differences in carbon-use efficiency (respiration : assimilation ratio). Other studies have demonstrated significant relationships between carbon-use efficiency and relative growth rates (Lambers and Rychter, 1989; Anekonda *et al.*, 1994). Several others have established links between what can be considered surrogates of carbon-use efficiency (cost versus benefit of foliage construction and maintenance) and plant carbon dynamics (Williams *et al.*, 1989; Sobrado, 1991; Griffin, 1994). Moreover, there is evidence that resource-use efficiencies decrease with increasing construction and maintenance costs but

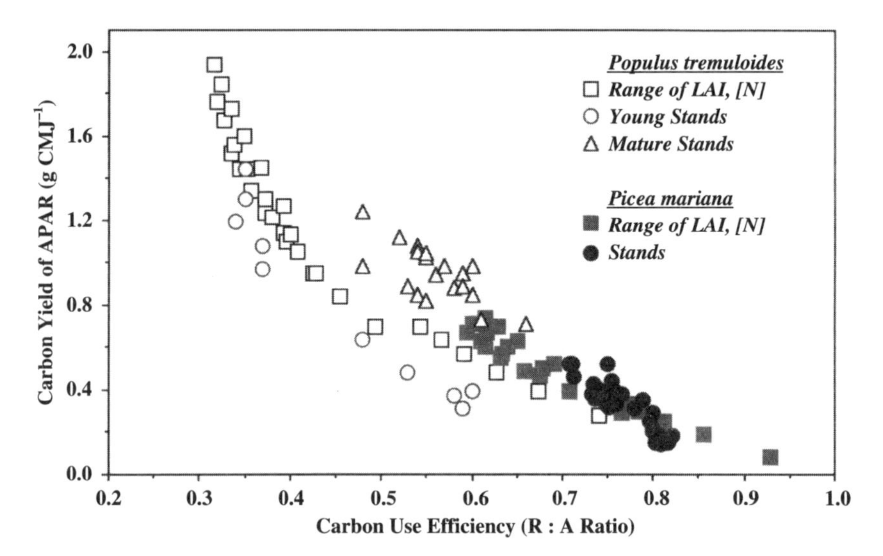

Fig. 4. Simulated annual carbon use efficiency (respiration : assimilation ratio) relative to annual carbon yield of APAR (i.e. APAR conversion to net primary production or ϵ_n) for two boreal tree species representing contrasted functional types. Each circle or triangle symbol represents an individual simulation parameterised with stand measurements (for each of 63 stands in northeast Minnesota). Each square symbol represents a simulation for a given LAI and foliage nitrogen [N] value, for a range of 1–5 LAI, 1.25–2.5% [N] for *Populus tremuloides* and 0.5–1.5% for *Picea mariana*. This figure was adapted from Goetz and Prince (1998). Individual stand measurements are reported by Goetz and Prince (1996).

that these increased costs are eventually recovered through greater foliage longevity (i.e. longer "payback interval") (Williams *et al.*, 1989; Griffin, 1994).

Calorimetric estimates of the costs of foliage construction are shown in Figure 5 for a range of different plant functional types. Note that for similar plant growth forms (shrubs, trees, etc.) in the same ecosystems (tropical dry forest, Mediterranean, etc.) species with greater leaf longevity (usually "evergreen") tend to have higher construction costs than species with shorter-lived leaves (deciduous). The sole exception to this is evergreen wet tropical shrubs, which tend to occur as understory plants in low-light environments. Such plants tend to have low leaf mass per unit area and shorter-lived leaves than many evergreen species, as well as unusual strategies of light absorption (red undersurfaces, epidermal lenses, blue iridescence), which affect their developmental control (Lee, 1986) and structural expense. Although not universal, and a topic of some debate (see Griffin, 1994), the association between construction cost and longevity is consistent with decreased carbon-use efficiency (as defined above) and extended payback intervals in species with longer-lived foliage. Foliage that is expensive to construct also tends to be expensive to maintain (Sobrado, 1994) but maintenance respiration costs differ in several ways among plant functional types, owing, for example, to

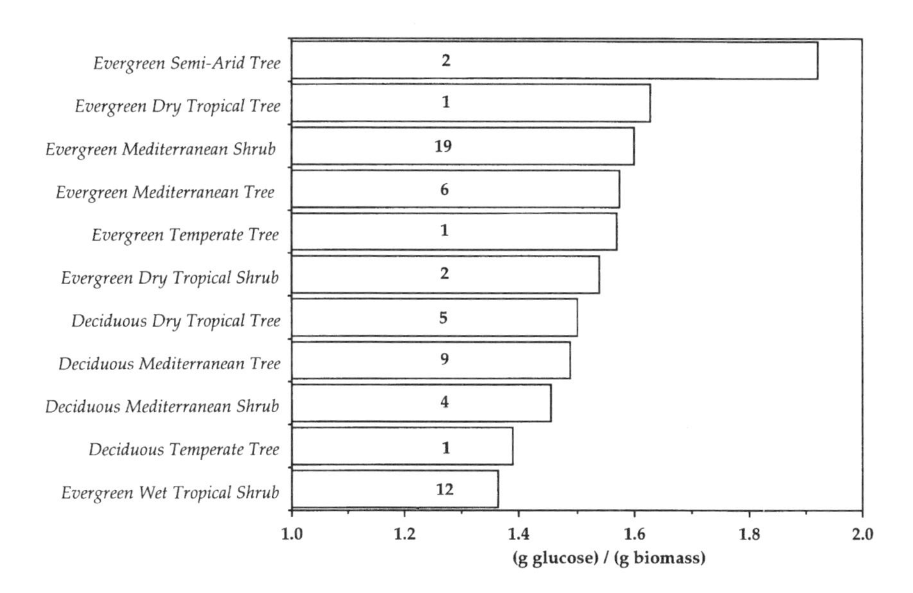

Fig. 5. Calorimetric foliage construction costs of a broad survey of plant functional types expressed in glucose units (data from Griffin, 1994). Numbers on each bar represent the sample size for that category.

differences in the proportion of respiring versus non-respiring (e.g. skeletal) plant components (Lambers and Poorter, 1992).

The association of differences in respiratory costs, foliage structure, longevity and, particularly, whole-plant biomass allocation and payback intervals among functional types is also evident at the stand level. For example, Goetz and Prince (1998) reported significantly lower carbon-use efficiencies for long-lived, slow-growing *Picea mariana* stands than for short-lived, fast-growing *Populus tremuloides* stands within the same ecosystem. Recent measurements in central Canada support the suggestion of a wide range in carbon-use efficiency among plant functional types (Ryan *et al.*, 1997). Carbon-use efficiency also varies within functional types; for example, a survey of the literature for complete stand carbon budgets of *Pinus* species in different environments revealed carbon-use efficiencies between 0.32 and 0.64 ($N = 7$) (Ryan *et al.*, 1994). Suggestions of a narrow range in carbon-use efficiency (at ~0.4) (Gifford, 1994; Landsberg *et al.*, 1997; Waring *et al.*, 1998) do not appear to apply to systems we have studied (e.g. boreal forests) despite positive correlations between respiration and assimilation (Amthor, 1995; Goetz and Prince, 1998). Extensive observations that maintenance respiration is proportional to respiring mass rather than total assimilation (e.g. Biscoe *et al.*, 1975; Ryan, 1990; Ryan *et al.*, 1997) support our contention that carbon-use efficiency varies primarily between different plant functional types in relation to resource allocation, defence costs and payback intervals on investment in carbon gain.

A common theme shared by studies of resource acquisition and allocation is that suites of life-history traits within functional types reflect an efficient allocation and use of available resources. Some have suggested the efficient allocation of resources has resulted in an optimisation of resource use such that fitness is maximised (Maynard Smith, 1978). Principles of natural selection are even applied to computational optimisation techniques (Forrest, 1993). The functional convergence hypothesis we explore here has its roots in optimality theory and so we next assess the evidence for maximisation of resource-use efficiency in plants.

VII. EVIDENCE FOR RESOURCE-USE EFFICIENCY

In conditions of sub-optimal resource supply, two types of response might be elicited in plants, one to increase the allocation of existing resources to the capture of the resource in short supply and the other to increase the efficiency of use of the resource in short supply. An overview follows of the allocation and use efficiencies of resources when in sub-optimal supply, considering nitrogen, water and light in turn, and finally the co-ordination of these.

A. Nitrogen

Hypotheses of optimised benefits relative to investment costs for nitrogen acquisition have been explored for more than two decades (see Gutschick, 1981). In a seminal study, Vitousek (1982) provided evidence that nutrient-use efficiency (NUE; carbon gain relative to nutrient use) in a range of mature forest ecosystems was inversely related to nutrient availability. Two mechanisms were advanced to explain this observation; increased resorption of nutrients from leaves to stems prior to abcision in nutrient poor sites, and increased NUE in photosynthetically active leaves.

There has been a plethora of studies that confirm and expand upon these observations. It has been suggested that, because resource allocation to construction of new foliage and uptake mechanisms is expensive for slow-growing species, high NUE is positively related to foliage longevity (Field, 1983; Reich *et al.*, 1992, 1997), the retention of nutrients in more productive foliage (Hom and Oechel, 1983; Hollinger, 1989), efficient nutrient resorption before shedding (Pugnaire and Chapin III, 1992), and reduced costs of nutrient storage (Chapin *et al.*, 1990; Monson *et al.*, 1994). The oldest foliage of black spruce stands in Alaska, for example, had nitrogen concentrations 70% less than the maximum values in new foliage (Hom and Oechel, 1983). A similar trend was observed with phosphorus, which was 55% less in old than in new foliage. Canopy nutrient distribution in relation to other factors (light and foliage mass; Figure 6) are discussed below.

Recent work with a large survey of tree species suggests that NUE is more similar between functional types of trees when expressed in terms of net production (i.e. time-integrated net photosynthesis) rather than the more commonly used ρ_c (Yin, 1994). This finding supports the contention that functional convergence should be assessed in the context of a time-averaged response to resource availability rather than the more adaptively plastic short-term response of ρ_c (which may vary, for example, with APAR; Björkman, 1981). Others (e.g. Field and Mooney, 1986; Gutschick, 1993; van der Werf *et al.*, 1993) have noted that variability in the relationship between N and CO_2 assimilation may be introduced by temporal changes in N-allocation patterns among and between functional types. These studies support the view that there is a time-integrated optimisation of N use, despite (or as a result of) dynamic allocation in plant tissues, for example, when resources are limiting. This dynamic optimisation process has also been incorporated into general ecophysiological models such that critical C : N ratios are maintained (Parton *et al.*, 1988; McGuire *et al.*, 1992; Bonan 1993).

B. Water

The trade-offs involved in optimising CO_2 uptake under conditions of water stress have been widely studied (Cowan, 1986; Givnish, 1986; Bégué *et al.*,

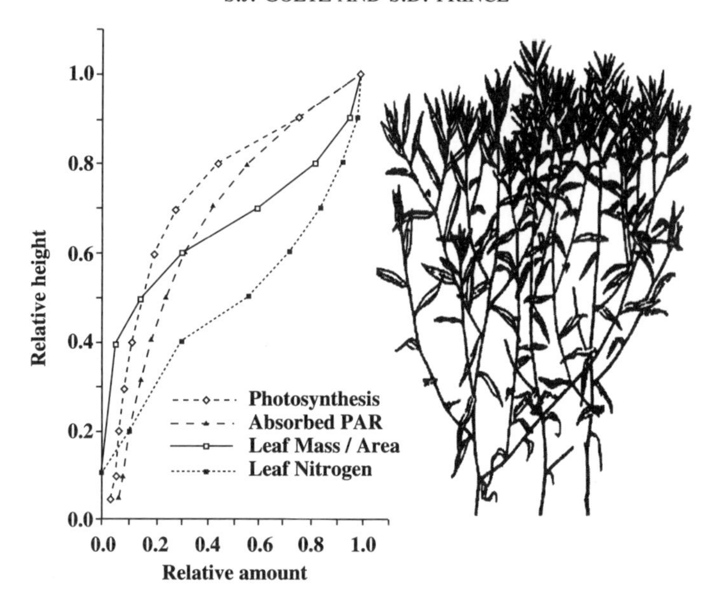

Fig. 6. Distribution of canopy nitrogen content, light absorption, leaf mass per unit area and photosynthesis with canopy height. Canopy profiles are based on actual observations of different species by Chen *et al.* (1993), Hirose and Werger (1987), Hollinger (1989), Hom and Oechel (1983) and Kull *et al.* (1995). Canopy diagram modified from figure of *Aster lanceolatus* (Schmidt and Bazzaz, 1994). – – ◇ – –, photosynthesis; – – ▲ – – , APAR; —□—, leaf mass/area; ···· ■ ····, leaf nitrogen.

1991) and expressed in a number of indices (Larcher, 1995). Of these indices, water-use efficiency (WUE) is the most relevant here as it expresses the ratio of dry matter production to water consumption by the plant or stand. Increased WUE in conditions of water shortage is well documented, as are the differences between species and plants having different carbon-fixation pathways (i.e. C3, C4, CAM) (Larcher, 1995). Mechanisms have been identified at all temporal scales from minutes to decades. Over longer time periods the emergence of different carbon-fixation pathways and their associated physiological and anatomical characteristics can be interpreted as an optimisation of carbon fixation per unit of available water. At shorter timescales, on the order of days, plants are known to reallocate carbon to the synthesis of osmotically active substances and hormones such as abscisic acid (ABA), which has a direct action on stomatal conductance and on carbon allocation to leaf growth, and to fine root growth (Geiger and Servaites, 1991). At the timescale of minutes, adjustments of leaf attitude, orientation, folding and curling may reduce the net radiation load (Forseth and Ehleringer, 1982; Ehleringer and Werk, 1986), thereby reducing water loss without much affecting carbon dioxide uptake and so optimising the WUE. These various responses to limited water give rise to the well-known differences in WUE between functional types. Abrams *et al.* (1994), for example, have linked WUE to leaf structural properties (leaf thickness and mass per unit area) and CO_2 exchange in a range of 17 broadleaf temperate tree species.

C. Photosynthetically Active Radiation (PAR)

Efficiency of PAR utilisation is a focus of this paper and is addressed primarily in the context of availability and utilisation of other resources. However, it is worth noting a few fundamental aspects of light utilisation. Optimisation of light capture relative to costs is supported by the evidence for resource "foraging" (Grime, 1979; Givnish, 1982) by means of increased leaf or height growth, more favourable positioning of leaves and changes in specific leaf area, all of which are associated with species that typically occupy partially shaded sites. Species that occupy sites in which the light environment may change rapidly from full sun to partial shade often have enhanced capability of modification of all these structural variables (Björkman, 1981; Walters and Field, 1987; Ellsworth and Reich, 1992). In contrast to allocation and structural adjustments, light-use efficiencies can change through time, mainly through changes in respiration with ontogeny (Perrin and Sibley, 1993; Goetz and Prince, 1996) or owing to environmental physiology (Potter *et al.*, 1993; Waring *et al.*, 1995; Goetz and Prince, 1998).

It is clear from these considerations of resource-use efficiency that the efficient use of any single resource (whether nutrients, water or light) is linked to the availability of all other resources, in much the same way that single plant traits are linked to the other traits associated with a given plant functional type. That is, various aspects of light-use efficiency are linked to the availability of other resources such that a combined response is elicited to any single resource that limits NPP, reproduction and growth.

VIII. CO-ORDINATED RESOURCE-USE EFFICIENCIES

The combined responses of plants to multiple limited resources requires an enhanced "law of the minimum" concept (Bloom *et al.*, 1985; Chapin *et al.*, 1987). A large number of studies in a diverse array of habitats have indicated, for example, that the distribution of N and the location of PAR absorption within plant canopies is such that both PAR and nitrogen concentration per unit leaf area decrease exponentially through the canopy (Figure 6). Linking the allocation of nitrogen through the canopy to light absorption results in high carbon gain and optimised resource use (Field, 1983; Ellsworth and Reich, 1993; Schmid and Bazzaz, 1994). This characteristic of efficient resource allocation has allowed mathematical formalisation of a PAR use parameter of canopy performance at all levels of a canopy to be scaled to that of the most illuminated leaves (Sellers *et al.*, 1992).

Such resource-allocation patterns, when combined with locally adapted canopy geometries (Horn, 1971; Gholz *et al.*, 1991; Farnsworth and Niklas, 1995), act to utilise the canopy light environment efficiently in relation to all other resources. In this way resource use is scaled to the limits of other resources available to the plant in a co-ordinated fashion (see Chapin *et al.*, 1987; Mooney *et al.*, 1991). Importantly, Lambers and Poorter (1992) reviewed plant responses to multiple stresses and found little evidence for trade-offs in resource-use efficiencies; the efficient utilisation of one resource was not compromised at the expense of the efficient utilisation of other resources.

Some researchers have thus preferred to refer to evidence for optimisation as "co-ordination" of resources (Chen *et al.*, 1993). Others have discussed it in the context of multiple evolutionarily stable strategies that achieve as close an optimal solution as genetic mechanisms permit (infinite time and infinite populations would be required to reach the theoretical optimum) (Parker and Maynard Smith, 1990). Debates over optimisation criteria do not detract from increasingly abundant evidence that the established linkages between traits, the trade-off costs associated with different allocation strategies and the response of plants to a variety of limited resources together support the idea that plant evolution has resulted in highly efficient and co-ordinated resource use. Because the optimisation debate has little direct relevance to our assessment of functional convergence, we have adopted co-ordination rather than optimisation terminology. A corollary to the evidence for co-ordination is that "wasteful allocation", such as unnecessary light harvesting and carbon-fixation machinery (defined in terms of the most limiting resource that prevents utilisation of additional carbon gain) is selected against and this is clearly in accordance with the functional convergence hypothesis. We emphasise once again that we refer to whole plant canopies rather than individual leaves or plant components.

It has been suggested that resource co-ordination resulting from natural selection can be expected to favour "not a maximum gross benefit but a compromise, optimum gross benefit which maximises the net benefit" (Begon *et al.*, 1986). The available evidence for resource-use efficiency supports this compromise interpretation. For example, Table 1 summarised the association between efficient allocation strategies, but also illustrates Reich *et al.*,'s (1992) conclusion regarding the co-ordination of resource use in fast-growing species: high leaf N concentrations result in higher CO_2 assimilation and greater canopy growth in proportion to total plant growth, which results in greater canopy expansion and height growth and increased canopy carbon gain. Increased carbon gain, in turn, results in high leaf turnover rates (short leaf lifespans) and associated rapid nutrient recycling. Slow-growing species are the reverse. Although this is a simplified scenario, it illustrates the mechanism by which co-ordinated resource-use efficiency, integrated through time, results in a maximised net benefit (i.e. carbon gain) for different plant functional types.

Co-ordination of multiple resource-use efficiencies may be expected to result in convergence on a narrow range of ϵ, as the functional convergence hypothesis predicts but, as noted earlier, such convergence is not evident in existing data sets (Figure 3). Lack of convergence may, however, be a consequence of how the results are compared, rather than a failure of the hypothesis, as the next section explores.

IX. LINKS BETWEEN CO-ORDINATED RESOURCE USE AND LEAF MASS PER UNIT AREA

Recent work suggests that the expression of energy use on a mass basis, as is the case in ϵ, is appropriate for comparisons of co-ordinated resource-use among plants. Several studies (Field and Mooney, 1986; Reich *et al.*, 1992, 1997; Agustí *et al.*, 1994) compared a wide variety of plant species that occur in a diverse range of environments and noted that resource-use efficiencies converge on a leaf-mass basis more so than on a leaf-area basis. For example, Field and Mooney (1986) noted the relationship between foliar nitrogen concentration and CO_2 assimilation was nearly linear when expressed on a leaf-mass basis, yet relatively poorly related on a leaf-area basis. Similarly, Reich *et al.* (1992) found NPP per unit foliage biomass varied by almost an order of magnitude and was closely related to suites of life-history traits (e.g. foliage longevity, $r^2 = 0.78$), whereas NPP on a unit leaf area basis was not significantly correlated with any of these same traits.

A mass basis for resource-use co-ordination emerges from associations between the expenditure of captured resources by the plant and the environmental resource distribution. Links between the canopy distribution of resources and the distribution of plant mass reflect the cost of acquisition, whereas areal metrics (e.g. LAI) reflect the amount of resource acquisition. In this way the cost of resource acquisition is best indicated by the distribution of leaf mass per unit area (LMA; reciprocal of specific leaf area, g m^{-2}). The evidence for efficient canopy N distribution, discussed above, has been shown to reflect this mass distribution, that is, nitrogen is distributed both within and among canopies (and differing light environment habitats) according to a scaling of LMA rather than leaf area (Walters and Field, 1987; Gutschick and Wiegel, 1988; Ellsworth and Reich, 1993) (Figure 6). We note that sub-optimal canopy distribution of N may occur in individual plants, owing to competition from adjacent individuals or differences in ontogeny (Schmid and Bazzaz, 1994). Nevertheless, current evidence suggests that resource-use co-ordination, resulting from combined resource-use efficiencies, is apparent only when the distribution of those resources is considered in the context of LMA.

Links between LMA and remote sensing of PAR absorption have been established and are explored below following an examination of the relevance of resource co-ordination and canopy carbon gain to fitness.

X. LINKS BETWEEN MAXIMISED CARBON GAIN AND FITNESS

Natural selection requires that co-ordinated resource-use results in maximisation of fitness and lifetime reproductive potential (Lovett Doust, 1989; Sibley, 1989; Burns, 1992). Does this translate into a maximisation of carbon gain, in terms of either NPP (growth) or gross primary productivity (total CO_2 assimilation), or is fitness unrelated to carbon gain?

Plants have a modular construction which allows them to be viewed as a "population of parts", in which plasticity is expressed not only in leaf shape and size but also in variation of the number of parts (Harper, 1990). More energy translates to more modules; more modules, more propagules; more propagules, more reproductive potential. The fate of assimilated carbon, however, is determined by the different allocation of resources to various plant processes, the trade-offs between resource allocation for growth versus non-growth (secondary) processes, and the different observed carbon-use efficiencies associated with plant functional types. Assimilation needs to be sufficient to provide the organic sugars for the various metabolic requirements of the plant, but that energy is not directly translated into biomass since other processes (such as defence) make different demands on energy from respiration of fixed carbon. Maximisation of carbon gain should confer fitness by providing the energy that can be allocated in any of a number of strategies that in turn maximise competitive advantage and reproductive potential. Genotypes that assimilate, allocate and utilise resources more efficiently than others that compete for the same resources would prevail, thereby resulting in a correlative relationship between co-ordinated resource use, maximised carbon gain and fitness (Parker and Maynard Smith, 1990; Sibley, 1991; Burns, 1992). We note that this observation does not suggest that natural selection acts largely on factors related to growth, rather it links the maximisation of carbon gain with traits that may favour reproductive potential through competitive advantage (i.e. fitness). This may be a consequence of past selection having eliminated most of the non-adaptive variation in characteristics related to growth (Field, personal communication).

These observations assume a long-term selection process under conditions in which vegetation may change with climate, environmental conditions, disturbance frequency and so on. The short-term competitive advantage of a species conferred by resource-use efficiency may be lost as conditions change or are reinitialised by disturbance events (e.g. Ehleringer, 1993; Perrin and Sibley, 1993). As noted earlier, such competitive advantage might only be conferred by more efficient resource use if this incorporated the combined utilisation of all resources. Moreover, these observations suggest that a mixture of functional types within an ecosystem provides the means by which a maximisation of carbon gain may occur at the level of the plot, stand or ecosystem rather than on an individual plant basis (e.g. Hirose and Werger, 1995). We are not suggesting group selection, we merely note that sessile individuals competing for resources in a given area may achieve maximised carbon gain as a result of their interaction and niche dynamics. That is, a variety of adaptations by a stand of competing individuals can result in maximised carbon gain, scaled to the available resources (e.g. Schmidt and Bazzaz, 1994). This also supports a tenet of the functional convergence hypothesis.

Although co-ordination of resource-use efficiency within available resource constraints may result in a maximisation of gross carbon gain, this will not necessarily result in a maximisation of net carbon gain (i.e. assimilation less respiratory costs) unless greater fitness or reproductive potential is somehow conferred. There is evidence to the contrary, that growth is not related to fitness (Sibley, 1989; Perrin and Sibley, 1993). Maximisation of NPP does not necessarily follow maximised carbon assimilation owing to differences in the relative energy (carbon) costs associated with various growth and allocation strategies. This is apparently true even within the same species; a study of bigtooth aspen (*Populus grandidentata*) stands over a range of site qualities demonstrated that CO_2 assimilation was positively correlated with NPP, but was not its sole determinant (Briggs *et al.*, 1986). A review by Körner (1991) describes additional cases where carbon assimilation may not provide a good prediction of NPP, and the findings of several studies with a large number of species (reviewed earlier) support this conclusion.

Theory and available evidence therefore support the view that there is, at best, a weak convergence of ϵ_n owing to differences in carbon-use efficiency associated with growth versus non-growth processes (e.g. Table 1, Figure 4) and associated "payback intervals" in different plant functional types. Instead, there is evidence for convergence of ϵ_g, which represents the gross assimilation of carbon rather than net production. There is also increasing evidence that ϵ_g* (the unstressed form of ϵ_g) approximates the quantum yield of photosynthesis (ϕ) (Prince and Goward, 1995; Waring *et al.*, 1995; Goetz and Prince, 1998), a term which reflects the full capacity at which the photosynthetic biochemical machinery may be expected to operate in order to offset respiratory costs sufficiently. This conclusion is further supported by field measurements of CO_2 exchange normalised by absorbed photosynthetically active radiation (APAR), which are similar among tropical, temperate and boreal ecosystems (Fan *et al.*, 1990; Ruimy *et al.*, 1995), although smaller than derived values of ϕ (e.g. Dewar, 1996; Haxeltine and Prentice, 1996; Goetz and Prince, 1998). Thus, we conclude that gross carbon assimilation and fitness are tightly linked, but the association between fitness and growth (NPP) is less direct, although positively correlated through their common link with maximised carbon assimilation and the costs and benefits of increased canopy development.

XI. IMPLICATIONS FOR REMOTE SENSING OF PRIMARY PRODUCTION

The evidence we have presented suggests that many of the vegetation processes and variables used in complex ecophysiology models that simulate carbon exchange and storage are not independent. Thus models that utilise many of these variables may be functionally simpler than stated. This has been demonstrated in exercises utilising functional-type classifications in ecophysiology models (Bonan, 1993; Bugmann, 1996), and provides credibility to models that reduce the dimensionality of autocorrelated variables and simplify vegetation into spatially uniform variables (characterised as "big leaf" approaches). Such functional simplications have also been used in stomatal conductance models (McNaughton and Jarvis, 1991).

Ecophysiology models, however, can be formulated on different bases to estimate carbon exchange over large areas. Several approaches have been explored, often providing similar results despite different relative importances ascribed to climate, vegetation type and resource availability (light, nutrients, etc.) (Field *et al.*, 1995; Hanan *et al.*, 1997). The main advantage of models that incorporate remotely sensed data is the spatially explicit monitoring capability provided by frequent observations of the land surface (Figure 2). Many global ecophysiology models now exploit remotely sensed data, at least to adjust parameter estimates and update state variables (e.g. Sellers *et al.*, 1992; Running and Hunt, 1993). The most sensitive components of terrestrial carbon models are frequently those which can be provided by remote sensing, that is, foliage dynamics (amount and longevity) and prominent physiological traits associated with functional-type classifications (see section V). This has led to the development of more simplified models driven predominantly by remotely sensed APAR (commonly referred to as production efficiency models; Prince *et al.*, 1994).

The association of plant canopy LMA with convergence in net photosynthesis among plant functional types (Reich *et al.*, 1992) is an important consideration for production efficiency models because spectral vegetation indices (SVIs) derived from remote sensing have been linked almost exclusively to leaf area (rather than mass) metrics. Recent work documenting high correlations between leaf mass, leaf optical properties and PAR absorption supports the relevance of LMA. Agustí *et al.* (1994), for example, found strong statistical relationships between the amount of light captured per unit weight of photosynthetic tissue, and both the chlorophyll concentration ($r^2 = 0.87$) and optical properties ($r^2 = 0.74$) of those tissues. Because of the nature of their comparisons among a diverse array of photosynthetic organisms, Agustí *et al.* (1994) did not examine leaf area metrics, however, Reich *et al.* (1997) have recently reiterated the importance of LMA as a fundamental variable in the global convergence of plant functioning (surveyed from 280 plant species around the globe). Additional research is clearly needed to determine the sensitivity of SVIs to LMA, which may improve not only the recovery of canopy state variables (LAI and biomass) but also those variables more representative of canopy process rates (F_{par}, ρ_c, ∇_f).

The second important consideration for primary production modelling with remote sensing is the difference in energy costs associated with construction of different plant components, and whether they are for growth or non-growth purposes. Remotely sensed recovery of APAR is intimately linked with GPP through the light reactions of photosynthesis but, because CO_2 respiration associated with growth processes differ, the use of APAR to estimate NPP requires some estimate of respiratory costs. This has been noted at a range of spatial scales (Running and Hunt, 1993; Goetz and Prince, 1996) and has been incorporated into production efficiency models that estimate NPP using remotely sensed data exclusively (Goward *et al.*, 1994; Prince and Goward, 1995). Moreover, the association between ϵ_g* and ϕ have allowed use of ϕ as a fundamental physiological constant from which to estimate ϵ_g by incorporating variables associated with short-term environmental stress (e.g. vapour pressure deficit, soil moisture) (Prince and Goward, 1995; Waring *et al.*, 1995). We believe functional convergence in light-use efficiency is, therefore, most relevant to primary production modelling through restrictions in the potential range of ϵ_g to a more conservative actual range.

The evidence for resource-use co-ordination, maximisation of biochemical capacity and convergence among functional types on a mass-per-unit-area basis suggests that "ecological integrators", particularly ϵ_g, can be used to simplify the biophysical and physiological components of models used to estimate net annual carbon fluxes over large areas. Despite early attempts to use SVIs alone, we believe remote sensing of NPP has little biophysical or physiological basis unless it incorporates stress and respiratory terms. Similarly, suggestions of NPP as a constant proportion of GPP (Landsberg and Waring, 1997; Waring *et al.*, 1998) are supported neither by the concept of payback intervals on resource investment (section VI) nor by recent carbon budget estimates for boreal forest stands (Gower *et al.*, 1997; Ryan *et al.*, 1997; Goetz and Prince, 1998). Simplification of the estimation of ϵ_g, how-ever, enables remote sensing to utilise a robust approach, based in evolutionary ecology, while exploiting the key advantage of a spatially and temporally contiguous monitoring capability.

XII. OVERVIEW AND CONCLUSIONS

Uncertainties in global carbon budgets can be reduced through improved characterisation of carbon exchange in terrestrial ecosystems. Satellite remote sensing, which has the distinct advantage of nearly continuous observations over large areas, offers one of several approaches to improve estimates of terrestrial primary production at a global scale. The task of accurately estimating canopy carbon exchange globally with remotely sensed data can be simplified if light-use efficiency of NPP (ϵ_n) varies little or at least predictably in different vegetation types. There have been many suggestions in the literature of a narrow range in ϵ_n, and much debate on its utility. It has been suggested that, owing to evolutionary co-ordination of resource-use efficiency, there may be an evolutionary basis for convergence in ϵ_n (Field, 1991).

The functional convergence hypothesis states that resource shortages of any sort will lead to adjustments of light capture, hence light capture serves as an integrator of resource status and biochemical capacity for CO_2 assimilation. We have examined the hypothesis in the context of remote sensing of terrestrial carbon exchange through an integration of inter-disciplinary research on resource-based growth constraints (stresses) and associated resource trade-offs, the costs versus benefits of various allocation strategies (particularly growth versus defence allocation), the evidence for co-ordination of resource-use efficiency, and the associated evidence for maximisation of carbon gain and fitness. Based on a synthesis of these several lines of research, we conclude that functional convergence is not apparent nor to be expected in ϵ_n, owing primarily to different payback intervals on carbon investment among different plant functional types. There are, however, ecological arguments and supporting evidence for functional convergence in the range of ϵ_g, owing to the efficient scaling of light-harvesting capacity to resource availability. These conclusions are elaborated below.

A preponderance of evidence suggests that natural selection results in closely related suites of traits which, acting together, improve fitness through the competitive advantage conferred by co-ordinated resource use. Several traits have been identified as "ecological integrators" that characterise functional types and their associated allocation of resources. Remotely sensed spectral vegetation indices are related to foliage display and light absorption by vegetation, and are thereby linked with many other plant traits. These traits reflect trade-offs in costs versus benefits of resource acquisition, allocation and utilisation. Plant functional types, defined by suites of related traits, appear to converge in their functioning (e.g. canopy process rates such as net photosynthesis) on a leaf-mass basis, rather than on a leaf area or volume basis, owing to these indices of resource acquisition.

The evidence for co-ordination of resource use suggests that a maximisation of carbon gain (not growth) is linked with a maximisation of fitness. The allocation of assimilated carbon, however, varies by growth strategy such that differences in respiratory costs relative to carbon gain do not result in a convergence on a narrow range in ϵ_n among plant species. For example, proportionately more resources are allocated to non-photosynthetic components in long-lived plant functional types adapted to resource-poor environments, which in turn results in greater respiratory costs relative to carbon gain. Thus decoupling of light harvesting and utilisation may occur via biochemical channels, which affects the linkages between maximised carbon assimilation, ϵ_n, and production models driven by light absorption.

Two important factors which may act to decouple light harvesting and carbon assimilation from carbon storage and ϵ_n have thus been confirmed: short-term stomatal control and longer-term respiratory carbon costs in relation to assimilatory carbon gains (carbon-use efficiency). Limitations in the convergence of light-use efficiency with respect to short-term decoupling of carbon assimilation and light harvesting has been noted, but longer-term decoupling associated with carbon-use efficiency has not. We have shown that ϵ_n varies among functional types primarily as a result of differences in carbon-use efficiency associated with different resource allocation strategies. This finding limits the direct applicability of functional convergence in light-use efficiency to the modelling of net primary production. We have instead found evidence for convergence in the amount of carbon assimilation per unit APAR (ϵ_g), and noted links between this term and the quantum yield of photosynthesis, a measure of maximum biochemical capacity for CO_2 assimilation. We therefore recommend that light-use efficiency, and its convergence, be applied to that aspect of terrestrial primary production modelling where it most benefits the utilisation of remotely sensed observations, that is, the determination of ϵ_g.

Our conclusions are based on a survey of the existing literature of plant ecophysiology, biophysics and evolutionary biology, all of which are rapidly expanding fields in global ecological research. They should be tested further with additional measurements of light absorption and whole-ecosystem production, as such data become available. It is particularly important to examine respiratory costs in relation to carbon gains for a range of resource allocation strategies, if accurate monitoring of terrestrial carbon exchange and storage with satellite remote sensing is to be realised.

ACKNOWLEDGMENTS

Chris Field is acknowledged for inspiring the inquiry. We thank those who reviewed the manuscript for their helpful comments. This work was partially supported by NASA grant NAGW 1967 to SDP.

REFERENCES

Abrams, M.D., Kubiske, M.E. and Mostoller, S.A. (1994). Relating wet and dry year ecophysiology to leaf structure in contrasting temperate tree species. *Ecology* **75**, 123–133.

Agustí, S., Enríquez, S., Frost-Christensen, H., Sand-Jensen, K. and Duarte, C.M. (1994). Light harvesting among photosynthetic organisms. *Functional Ecology* **8**, 273–279.

Amthor, J.S. (1995). Higher plant respiration and its relationships to photosynthesis. In: *Ecophysiology of Photosynthesis* (Ed. by E.-D. Schulze and M.M. Caldwell), pp. 71–101. Springer Verlag, Berlin.

Anekonda, T.S., Criddle, R.S., Libby, W.J., Breidenbach, R.W. and Hansen, L.D. (1994). Respiration rates predict differences in growth of coast redwood. *Plant, Cell and Environment* **17**, 197–203.

Asrar, G., Fuchs, M., Kanemasu, E.T. and Hatfield, J.L. (1984). Estimating absorbed photosynthetic radiation and leaf area index from spectral reflectance in wheat. *Agronomy Journal* **76**, 300–306.

Asrar, G., Kanemasu, E.T., Jackson, R.D. and Pinter, P.J. (1985). Estimation of total above ground phytomass production using remotely sensed data. *Remote Sensing of Environment* **17**, 211–220.

Bazilevich, N.I., Rodin, L.Y. and Rozov, N.N. (1971). Geographical aspects of biological productivity. *Soviet Geography* **12**, 293–317.

Begon, M., Harper, J.L. and Townsend, C.R. (1986). *Ecology*. Blackwell Scientific, Oxford.

Bégué, A., Desprat, J., Imbernon, J. and Baret, F. (1991). Radiation use efficiency of pearl millet in the Sahelian zone. *Agricultural and Forest Meteorology* **56**, 93–110.

Bégué, A., Roujean, J.L., Hanan, N.P., Prince, S.D., Thawley, M., Huete, A. and Tanré, D. (1996). Shortwave radiation budget of Sahelian vegetation: Techniques of measurement and results during HAPEX-Sahel. *Agricultural and Forest Meteorology* **79**, 79–96.

Biscoe, P.V., Scott, R.K. and Monteith, J.L. (1975). Barley and its environment III. Carbon budget of the stand. *Journal of Applied Ecology* **12**, 269–293.

Björkman, O. (1981). Responses to different quantum flux densities. In: *Physiological Plant Ecology* (Ed. by O.L. Lange, P.S. Nobel, C.B. Osmond and H. Ziegler), pp. 57–108. Springer Verlag, Berlin.

Bloom, A., Chapin III, F.S. and Mooney, H. (1985). Resource limitation in plants—an economic analogy. *Annual Review Ecology and Systematics* **16**, 363–392.

Bonan, G.B. (1993). Importance of leaf area index and forest type when estimating photosynthesis in boreal forests. *Remote Sensing of Environment* **43**, 303–314.

Box, E.O. (1996). Plant functional types and climate at the global scale. *Journal of Vegetation Science* **7**, 309–320.

Briggs, G.M., Jurik, T.W. and Gates, D.M. (1986). A comparison of rates of aboveground growth and carbon dioxide assimilation by aspen on sites of high and low quality. *Tree Physiology* **2**, 29–34.

Bugmann, H. (1996). Functional types of trees in temperate and boreal forests: classification and testing. *Journal of Vegetation Science* **7**, 359–370.

Burns, T.P. (1992). Adaptedness, evolution and a hierarchical concept of fitness. *Journal of Theoretical Biology* **154**, 219–237.

Cannell, M.G.R., Milne, R., Sheppard, L.J. and Unsworth, M.H. (1987). Radiation interception and productivity in willow. *Journal of Applied Ecology* **24**, 261–278.

Chabot, B.F. and Hicks, D.J. (1982). The ecology of leaf life-spans. *Annual Review of Ecology and Systematics* **13**, 229–259.

Chapin, F.S. (1993). Functional role of growth forms in ecosystem and global processes. In: *Scaling Physiological Processes* (Ed. by J.R. Ehleringer and C.B. Field), pp. 287–312. Academic Press, San Diego.
Chapin, F.S., Bloom, A.J., Field, C.B. and Waring, R.H. (1987). Plant responses to multiple environmental factors. *BioScience* **37**, 49–57.
Chapin, F.S., Schulze, E.-D. and Mooney, H.A. (1990). The ecology and economics of storage in plants. *Annual Review of Ecology and Systematics* **21**, 423–447.
Chen, J.-L., Reynolds, J.F., Harley, P.C. and Tenhunen, J.D. (1993). Co-ordination theory of leaf nitrogen distribution in a canopy. *Oecologia* **93**, 63–69.
Choudhury, B.J. (1987). Relationships between vegetation indices, radiation absorption, and net photosynthesis evaluated by a sensitivity analysis. *Remote Sensing of Environment* **22**, 209–233.
Ciais, P., Tans, P.P., Trolier, M., White, J.W.C. and Francey, R.J. (1995). A large northern hemisphere terrestrial CO_2 sink indicated by the $^{13}C/^{12}C$ ratio of atmospheric CO_2. *Science* **269**, 1098–1101.
Coley, P.D. (1986). Costs and benefits of defense by tannins in a neotropical tree. *Oecologia* **70**, 238–241.
Coley, P.D., Bryant, J.P. and Chapin, F.S. (1985). Resource availability and plant anti-herbivore defense. *Science* **230**, 895–899.
Collatz, G.J., Ball, J.T., Cirivet, C. and Berry, J.A. (1991). Physiological and environmental regulation of stomatal conductance, photosynthesis and transpiration: a model that includes a laminar boundary layer. *Agricultural and Forest Meteorology* **54**, 107–136.
Cowan, J.R. (1986). Economics of carbon fixation in higher plants. In: *On the Economics of Plant Form and Function* (Ed. by T.J. Givnish). Cambridge University Press, Cambridge.
Curran, P.J. (1983). Multispectral remote sensing for the estimation of green leaf area index. *Philosophical Transactions of the Royal Society of London A* **309**, 257–270.
Daughtry, C.S.T., Gallo, K.P., Goward, S.N., Prince, S.D. and Kustas, W.P. (1992). Spectral estimates of absorbed radiation and phytomass production in corn and soybean canopies. *Remote Sensing of Environment* **39**, 141–152.
Defries, R., Field, C.B., Fung, I., Justice, C.O., Los, S., Matson, P.A., Mooney, E.A., Potter, C.S., Prentice, K.A., Sellers, P.J., Townshend, J.R.G., Tucker, C.J., Ustin, S. and Vitousek, P.M. (1995). Mapping the land surface for global atmosphere-biosphere models: towards continuous distributions of vegetation's functional properties. *Journal of Geophysical Research* **100**, 20867–20882.
Denning, A.S., Fung, I.Y. and Randall, D. (1995). Latitudinal gradient of atmospheric CO_2 due to seasonal exchange with land biota. *Nature* **376**, 240–243.
Dewar, R.C. (1996). The correlation between plant growth and intercepted radiation: an interpretation in terms of optimal plant nitrogen content. *Annals of Botany* **78**, 125.
Dye, D. and Shibasaki, R. (1995). Intercomparison of global PAR data sets. *Geophysical Research Letters* **22**, 2013–2016.
Ehleringer, J.R. (1993). Variation in leaf carbon isotope discrimination in *Encilia farinosa*: implications for growth, competition and drought survival. *Oecologia* **95**, 340–355.
Ehleringer, J.H. and Werk, K.S. (1986). Modifications of solar radiation absorption in patterns and implications for carbon gain at the leaf level. In: *On the Economics of Plant Form and Function* (Ed. by T J. Givnish), pp. 57–82. Cambridge University Press, Cambridge.
Ellsworth, D.S. and Reich, P.B. (1992). Leaf mass per area, nitrogen content and photosynthetic carbon gain in *Acer saccharum* seedlings in contrasting forest light environments. *Functional Ecology* **6**, 423–435.

Ellsworth, D.S. and Reich, P.B. (1993). Canopy structure and vertical patterns of photosynthesis and related leaf traits in a deciduous forest. *Oecologia* **96**, 169–178.
Esser, G. (1991). Osnabrück Biosphere Model: structure, construction, results. In: *Modern Ecology: Basic and Applied Aspects* (Ed. by G. Esser and D. Overdieck). Elsevier, Amsterdam.
Fagerström, T., Larsson, S. and Tenow, O. (1987). On optimal defence in plants. *Functional Ecology* **1**, 73–81.
Fan, S.-M., Wofsy, S.C., Bakwin, P.S., Jacob, D.J. and Fitzjarrald, D.R. (1990). Atmosphere-biosphere exchange of CO_2 and O_3 in the central Amazon forest. *Journal of Geophysical Research* **95**, 16851–16864.
Farnsworth, K.D. and Niklas, K.J. (1995). Theories of optimization, form and function in branching architecture in plants. *Functional Ecology* **9**, 355–363.
Field, C.B. (1983). Allocating leaf nitrogen for the maximization of carbon gain: leaf age as a control on the allocation program. *Oecologia* **56**, 341–347.
Field, C.B. (1991). Ecological scaling of carbon gain to stress and resource availability. In: *Response of Plants to Multiple Stresses* (Ed. by H.A. Mooney, W.E. Winner and E.J. Pell), pp. 35–65. Academic Press, San Diego.
Field, C.B. and Mooney, H.A. (1986). The photosynthesis–nitrogen relationship in wild plants. In: *On the Economy of Plant Form and Function* (Ed. by T.J. Givnish), pp. 681–698. Cambridge University Press, Cambridge.
Field, C.B., Randerson, J.T. and Malmström, C.M. (1995). Global net primary production: combining ecology and remote sensing. *Remote Sensing of Environment* **51**, 74–88.
Foley, J.A., Prentice, I.C. and Haxeltine, A. (1996). An integrated biosphere model of land surface processes, terrestrial carbon balance, and vegetation dynamics. *Global Biogeochemical Cycles* **10**, 603–628.
Forrest, S. (1993). Genetic algorithms—principles of natural selection applied to computation. *Science* **261**, 872–878.
Forseth, I.N. and Ehleringer, J.R. (1982). Ecophysiology of two solar tracking desert annuals 2. leaf movements, water relations and microclimate. *Oecologia* **54**, 41–49.
Gadgil, M. and Bossert, W. (1970). Life history consequences of natural selection. *American Naturalist* **104**, 1–24.
Geiger, D.R. and Servaites, J.C. (1991). Carbon allocation and response to stress. In: *Response of Plants to Multiple Stress* (Ed. by H.A. Mooney, W.E. Winner and E.J. Pell), pp. 103–127. Academic Press, San Diego.
Gholtz, H.L., Vogel, S.A., Cropper, W.P., McKelvey, K., Ewel, K.C., Tesky, R.O. and Curran, P.J. (1991). Dynamics of canopy structure and light interception in *Pinus elliottii* stands, north Florida. *Ecological Monographs* **61**, 33–51.
Gifford, R.M. (1994). The global carbon cycle: a viewpoint on the missing sink. *Australian Journal of Plant Physiology* **21**, 1–15.
Givnish, T.J. (1982). On the adaptive significance of leaf height in forest herbs. *American Naturalist* **120**, 353–381.
Givnish, T.J. (1986). *On the Economy of Plant Form and Function*. Cambridge University Press, Cambridge.
Gleeson, S.K. and Tilman, D. (1994). Plant allocation, growth rate and successional status. *Functional Ecology* **8**, 543–550.
Goel, N. and Qin, W. (1994). Influence of canopy architecture on various vegetation indices and LAI and Fpar: simulation model results. *Remote Sensing Reviews* **10**, 309–347.
Goetz, S.J. (1997). Multi-sensor analysis of NDVI, surface temperature, and biophysical variables at a mixed grassland site. *International Journal of Remote Sensing* **18**, 71–94.

Goetz, S.J. and Prince, S.D. (1996). Remote sensing of net primary production in boreal forest stands. *Agricultural and Forest Meteorology* **78**, 149–179.

Goetz, S.J. and Prince, S.D. (1998). Variability in carbon exchange and light utilization among boreal forest stands: implications for remote sensing of net primary production. *Canadian Journal of Forest Research* **28**, 375–389.

Goward, S.N., Cruickshanks, G.D. and Hope, A.S. (1985). Observed relation between thermal emission and reflected spectral radiance of a complex vegetated landscape. *Remote Sensing of Environment* **18**, 137–146.

Goward, S.N. and Huemmrich, K.F. (1992). Vegetation canopy PAR absorbance and the normalized difference vegetation index: an assessment using the SAIL model. *Remote Sensing of Environment* **39**, 119–140.

Goward, S.N. and Prince, S.D. (1995). Transient effects of climate on vegetation dynamics: satellite observations. *Journal of Biogeography* **22**, 549–563.

Goward, S.N., Waring, R.H. and Dye, D.G. (1994). Ecological remote sensing at OTTER: Macroscale satellite observations. *Ecological Applications* **4**, 322–343.

Gower, S.T., Vogel, J., Norman, J., Kulharik, C., Steele, S. and Stow, T. (1997). Carbon distribution and above-ground net primary production of upland and lowland boreal forests in Saskatchewan and Manitoba, Canada. *Journal of Geophysical Research* **102**, 29029–29042.

Griffin, K.L. (1994). Calorimetric estimates of construction cost and their use in ecological studies. *Functional Ecology* **8**, 551–562.

Grime, J.P. (1979). *Plant Strategies and Vegetation Processes*. John Wiley and Sons, London.

Gutschick, V.P. (1981). Evolved strategies in nitrogen acquisition by plants. *American Naturalist* **118**, 607–637.

Gutschick, V.P. (1993). Nutrient-limited growth rates: Rules of nutrient-use efficiency and of adaptations to increased uptake rate. *Journal of Experimental Botany* **44**, 41–51.

Gutschick, V.P. and Wiegel, F.W. (1988). Optimizing the canopy photosynthetic rate by patterns of investment in specific leaf mass. *American Naturalist* **132**, 67–86.

Hanan, N.P., Prince, S.D. and Bégué, A. (1997). Modelling vegetation primary production during HAPEX-Sahel using production efficiency and canopy conductance model formulations. *Journal of Hydrology* **188/189**, 651–675.

Harper, J.L. (1989). The value of a leaf. *Oecologia* **80**, 53–58.

Harper, J.L. (1990). Canopies as populations. In: *Plant Canopies: Their Growth, Form and Function* (Ed. by G. Russel, B. Marshall and P.G. Jarvis), pp. 105–128. Cambridge University Press, Cambridge.

Haxeltine, A. and Prentice, I.C. (1996). A general model for the light-use efficiency of primary production. *Functional Ecology* **10**, 551–561.

Hirose, T. and Werger, M.J.A. (1987). Maximizing daily canopy photosynthesis with respect to the leaf nitrogen allocation pattern in the canopy. *Oecologia* **72**, 520–526.

Hirose, T. and Werger, M.J.A. (1995). Canopy structure and photon flux partitioning among species in a herbaceous plant community. *Ecology* **76**, 466–474.

Hollinger, D.Y. (1989). Canopy organization and foliage photosynthetic capacity in a broad-leaved evergreen montane forest. *Functional Ecology* **3**, 53–62.

Hom, J.L. and Oechel, W.C. (1983). The photosynthetic capacity, nutrient content, and nutrient use efficiency of different needle age-classes of black spruce (*Picea mariana*) found in interior Alaska. *Canadian Journal of Forest Research* **13**, 834–839.

Horn, H. (1971). *The Adaptive Geometry of Trees*. Princeton University Press, Princeton, NJ.

Hunt, E.R. and Running, S.W. (1992). Simulated dry matter yields for aspen and spruce stands in the North American boreal forest. *Canadian Journal of Remote Sensing* **18**, 126–133.

Hunt, E.R., Piper, S.C. and Running, S.W. (1996). Global net carbon exchange and intra-annual atmospheric CO_2 concentrations predicted by an ecosystem process model and three-dimensional atmospheric transport model. *Global Biogeochemical Cycles* **10**, 431–456.

IPCC (1996). *Climate Change 1995: The Science of Climate Change. Contribution of Working Group I to the Second Assessment Report of the Intergovernmental Panel on Climate Change* (Ed. by J.J. Houghton, L.G. Meiro, B.A. Filho, N. Callander, A. Harris, A. Kattenberg and K. Maskel). Cambridge University Press, Cambridge.

Kachi, N. and Rorison, I.H. (1989). Optimal partitioning between root and shoot in plants with contrasted growth rates in response to nitrogen availability and temperature. *Functional Ecology* **3**, 549–559.

Kawecki, T.J. (1993). Age and size at maturity in a patchy environment—fitness maximization versus evolutionary stability. *Oikos* **66**, 309–317.

Keeling, C.D., Whorf, T.P., Wahlen, M. and van der Plicht, J. (1995). Interannual extremes in the rate of rise of atmospheric carbon dioxide since 1980. *Nature* **375**, 666–670.

Keeling, R.F., Piper, S.C. and Heimann, M. (1996). Global and hemispheric CO_2 sinks deduced from changes in atmospheric O_2 concentration. *Nature* **381**, 218.

Knorr, W. and Heimann, M. (1996). Impact of drought stress and other factors on seasonal land biosphere CO_2 exchange studied through an atmospheric tracer transport model. *Tellus* **47**, 471–489.

Körner, C. (1991). Some often overlooked plant characteristics as determinants of plant growth: A reconsideration. *Functional Ecology* **5**, 162–173.

Kull, O., Aan, A. and Sõelsepp, T. (1995). Light interception, nitrogen and leaf mass distribution in a multilayer plant community. *Functional Ecology* **9**, 589–595.

Kumar, M. and Monteith, J.L. (1982). Remote sensing of plant growth. In: *Plants and the Daylight Spectrum* (Ed. by H. Smith), pp. 133–144. Academic Press, London.

Lambers, H. and Poorter, H. (1992). Inherent variation in growth rate between higher plants: A search for physiological causes and ecological consequences. *Advances in Ecological Research* **23**, 187–261.

Lambers, H. and Rychter, A. (1989). The biochemical background of variation in respiration rate: respiratory pathways and chemical composition. In: *Causes and Consequences of Variation in Growth Rate and Productivity of Higher Plants* (Ed. by H. Lambers, M. Cambridge, H. Konings and T.L. Pons), pp. 199–225. Academic Publishing, The Hague.

Landsberg, J.J. (1986). *Physiological Ecology of Forest Production*. Academic Press, London.

Landsberg, J.J., Prince, S.D., Jarvis, P.G., McMurtrie, R.E., Luxmore, R. and Medlyn, B.E. (1997). Energy conversion and use in forests: The analysis of forest production in terms of utilisation efficiency (ϵ). In: *Use of Remote Sensing in the Modeling of Forest Productivity* (Ed. by H.L. Gholtz, K. Nakane and H. Shimoda), pp. 273–298. Kluwer Academic Publishers, New York.

Landsberg, J.J. and Waring, R.H. (1997). A generalised model of forest productivity using simplified concepts of radiation-use efficiency, carbon balance and partitioning. *Forest Ecology and Management* **95**, 209–228.

Larcher, W. (1995). *Physiological Plant Ecology*. Springer Verlag, Heidelberg.

Laurence, J.A., Amundson, R.G., Friend, A.L., Pell, E.J. and Temple, P.J. (1994). Allocation of carbon in plants under stress: an analysis of the ROPIS experiments. *Journal of Environmental Quality* **23**, 412–417.

Lee, D.W. (1986). Unusual strategies of light absorption in rain-forest herbs. In: *On the Economy of Plant Form and Function* (Ed. by J.J. Givnish), pp. 105–132. Cambridge University Press, Cambridge.
Leith, H. and Box, E. (1977). The gross primary production pattern of the land vegetation: a first attempt. *Tropical Ecology* **18**, 109–115.
Lerdau, M. (1992). Future discounts and resource allocation in plants. *Functional Ecology* **6**, 371–375.
Lerdau, M., Litvak, M. and Monson, R. (1994). Plant chemical defense: monoterpenes and the growth-differentiation balance hypothesis. *Trends in Ecology and Evolution* **9**, 58–61.
Linder, S. (1985). Potential and actual production in Australian forest stands. In: *Research for Forest Management* (Ed. by J.J. Landsberg and W. Parsons), pp. 11–35. CSIRO, Melbourne.
Lloyd, J. and Farquhar, G.D. (1996). The CO_2 dependence of photosynthesis, plant growth responses to elevated atmospheric CO_2 concentrations and their interaction with soil nutrient status. I. General principles and forest ecosystems. *Functional Ecology* **10**, 4–32.
Loehle, C. (1988). Tree life history strategies: the role of defenses. *Canadian Journal of Forest Research* **18**, 209–222.
Lovett Doust, J. (1989). Plant reproductive strategies and resource allocation. *Trends in Ecology and Evolution* **4**, 230–234.
MacArthur, R.H. and Wilson, E.O. (1967). *The Theory of Island Biogeography.* Princeton University Press, Princeton.
McGuire, A.D., Melillo, J.M., Joyce, L.A., Kicklighter, D.W., Grace, A.L., Moore, B. and Vorosmarty, C.J. (1992). Interactions between carbon and nitrogen dynamics in estimating net primary production for potential vegetation in North America. *Global Biogeochemical Cycles* **6**, 101–124.
McNaughton, K.G. and Jarvis, P.G. (1991). Effects of spatial scale on stomatal control of transpiration. *Agricultural and Forest Meteorology* **54**, 279–301.
Maynard Smith, J. (1978). Optimization theory in evolution. *Annual Review Ecological Systematics* **9**, 31–56.
Melillo, J.M., McGuire, A.D., Kicklighter, D.W., Moore, B., Vorosmarty, C.J. and Schloss, A.L. (1993). Global climate change and terrestrial net primary production. *Nature* **363**, 234–240.
Monson, R.K., Schulze, E.-D., Freund, M. and Heilmeier, H. (1994). The influence of nitrogen availability on carbon and nitrogen storage in biennial *Cirsium vulgare (Savi)* II. The cost of nitrogen storage. *Plant, Cell and Environment* **17**, 1133–1141.
Monteith, J.L. (1977). Climate and the efficiency of crop production in Britain. *Philosophical Transactions of the Royal Society of London B* **281**, 277–294.
Monteith, J.L. (1994). Validity of the correlation between intercepted radiation and biomass. *Agricultural and Forest Meteorology* **68**, 213–220.
Mooney, H.A., Winner, W.E. and Pell, E.J. (1991). *Response of Plants to Multiple Stresses.* Academic Press, San Diego.
Nemani, R. and Running, S. (1997). Land cover characterization using multitemporal red, near-IR, and thermal-IR data from NOAA/AVHRR. *Ecological Applications* **7**, 79–90.
Oliver, C.D. and Larson, B.C. (1990). *Forest Stand Dynamics.* McGraw-Hill, Inc, New York.
Parker, G.A. and Maynard Smith, J. (1990). Optimality theory in evolutionary biology. *Nature* **348**, 27–33.
Parton, W.J., Stewart, J.W.B. and Cole, C.V. (1988). Dynamics of C, N, P, and S in grassland soils: a model. *Biogeochemistry* **5**, 109–131.

Partridge, L. and Harvey, P.H. (1988). The ecological context of life history evolution. *Science* **241**, 1449–1455.

Penning de Vries, F.W.T. (1975). The cost of maintenance processes in plant cells. *Annals of Botany* **39**, 77–92.

Perrin, N. and Sibley, R.M. (1993). Dynamic models of energy allocation and investment. *Annual Review of Ecology and Systematics* **24**, 379–410.

Potter, C.S., Randerson, J.T., Field, C.B., Matson, P.A., Vitousek, P.M., Mooney, H.A. and Klooster, S.A. (1993). Terrestrial ecosystem production: A process model based on global satellite and surface data. *Global Biogeochemical Cycles* **7**, 811–841.

Prince, S.D. (1991a). A model of regional primary production for use with coarse-resolution satellite data. *International Journal of Remote Sensing* **12**, 1313–1330.

Prince, S.D. (1991b). Satellite remote sensing of primary production: comparison of results for Sahelian grasslands 1981–1988. *International Journal of Remote Sensing* **12**, 1301–1330.

Prince, S.D. and Astle, W.L. (1986). Satellite remote sensing of rangelands in Botswana. I. Landsat MSS and herbaceous vegetation. *International Journal of Remote Sensing* **7**, 1533–1553.

Prince, S.D. and Goward, S.J. (1995). Global primary production: A remote sensing approach. *Journal of Biogeography* **22**, 815–835.

Prince, S.D., Justice, C.O. and Moore, B. (1994). *Monitoring and Modeling of Terrestrial Net and Gross Primary Production.* Working Paper 1, International Geosphere Biosphere Program (IGBP) Data and Information System (DIS) Global Analysis, Interpretation and Modeling (GAIM), 57 pp.

Pugnaire, F.I. and Chapin III, F.S. (1992). Environmental and physiological factors governing nutrient resorption efficiency in barley. *Oecologia* **90**, 120–126.

Rauner, J.L. (1976). Deciduous Forests. In: *Vegetation and the Atmosphere* (Ed. by J.L. Monteith), pp. 241–264. Academic Press, New York.

Raven, J.A. (1986). Evolution of plant life forms. In: *On the Economy of Plant Form and Function* (Ed. by T.J. Givnish), pp. 421–492. Cambridge University Press, London.

Reich, P.B., Walters, M.B. and Ellsworth, D.S. (1992). Leaf life-span in relation to leaf, plant, and stand characteristics among diverse ecosystems. *Ecological Monographs* **62**, 365–392.

Reich, P.B., Walters, M.B. and Ellsworth, D.S. (1997). From tropics to tundra: global convergence in plant functioning. *Proceedings of the National Academy of Sciences* **94**, 13730–13734.

Ricklefs, R.E. (1991). Structures and transformations of life histories. *Functional Ecology* **5**, 174–183.

Ruimy, A., Dedieu, G. and Saugier, B. (1994). Methodology for the estimation of terrestrial net primary production from remotely sensed data. *Journal of Geophysical Research* **99**, 5263–5283.

Ruimy, A., Jarvis, P., Baldocchi, D. and Saugier, B. (1995). CO_2 fluxes over plant canopies and solar radiation: a review. *Advances in Ecological Research* **26**, 1–51.

Running, S.W. and Hunt, E.R. (1993). Generalization of a forest ecosystem process model for other biomes, BIOME-BGC, and an application for global-scale models. In: *Scaling Physiological Processes* (Ed. by J.R. Ehleringer and C.B. Field), pp. 141–158. Academic Press, San Diego.

Running, S.W. and Nemani, R.R. (1987). Relating seasonal patterns of the AVHRR vegetation index to simulated photosynthesis and transpiration of forests in different climates. *Remote Sensing of Environment* **24**, 347–367.

Runyon, J., Waring, R.H., Goward, S.N. and Welles, J.M. (1994). Environmental limits on net primary production and light use efficiency across the Oregon transect. *Ecological Applications* **4**, 226–238.
Russel, G., Marshall, B. and Jarvis, P.G. (1989). *Plant Canopies: Their Growth, Form and Function.* Cambridge University Press, Cambridge.
Ryan, M.G. (1990). Growth and maintenance respiration in stems of *Pinus contorta* and *Picea engelmannii*. *Canadian Journal of Forest Research* **20**, 48–57.
Ryan, M.G., Linder, S., Vose, J.M. and Hubbard, R.M. (1994). Dark respiration of pines. *Ecological Bulletins* **43**, 50–63.
Ryan, M.G., Lavigne, M.B. and Gower, S.T. (1997). Annual carbon cost of autotrophic respiration in boreal forest ecosystems in relation to species and climate. *Journal of Geophysical Research* **102**, 28871–28884.
Saldarriaga, J.G. and Luxmoore, R.J. (1991). Solar energy conversion efficiencies during succession of a tropical rain forest in Amazonia. *Journal of Tropical Ecology* **7**, 233–242.
Schmid, B. and Bazzaz, F.A. (1994). Crown construction, leaf dynamics and carbon gain in two perennials with contrasting architecture. *Ecological Monographs* **64**, 177–203.
Sellers, P.J. (1985). Canopy reflectance, photosynthesis and transpiration. *International Journal of Remote Sensing* **6**, 1335–1372.
Sellers, P.J. (1987). Canopy reflectance, photosynthesis and transpiration. II. The role of biophysics in the linearity of their interdependence. *Remote Sensing of Environment* **21**, 143–183.
Sellers, P.J., Berry, J.A., Collatz, G.J., Field, C.B. and Hall, F.G. (1992). Canopy reflectance, photosynthesis and transpiration III: A reanalysis using improved leaf models and a new canopy integration scheme. *Remote Sensing of Environment* **42**, 187–216.
Sellers, P.J., Heiser, M.D., Hall, F.G., Goetz, S.J., Strebel, D.E., Verma, S.B., Dejardins, R.L., Schuepp, P.M. and MacPherson, J.I. (1995). The effects of spatial variability in topography, vegetation cover and soil moisture on area-averaged surface fluxes: A case study using the FIFE-89 data. *Journal of Geophysical Research* **100**, 25607–25630.
Shipley, B. and Peters, R.H. (1990). A test of the Tilman model of plant strategies: relative growth rate and biomass partitioning. *American Naturalist* **136**, 139–153.
Sibley, R.M. (1989). What evolution maximizes. *Functional Ecology* **3**, 129–135.
Sibley, R.M. (1991). The life-history approach to physiological ecology. *Functional Ecology* **5**, 184–191.
Silvertown, J., Franco, M. and McConway, K. (1992). A demographic interpretation of Grime's triangle. *Functional Ecology* **6**, 130–136.
Smith, T.M., Shugart, H.H. and Woodward, F.I. (1997). *Plant Functional Types: Their Relevance to Ecosystem Properties and Global Change*. Cambridge University Press, Cambridge.
Sobrado, M.A. (1991). Cost–benefit relationships in deciduous and evergreen leaves of tropical dry forest species. *Functional Ecology* **5**, 608–616.
Sobrado, M.A. (1994). Leaf age effects on photosynthetic rate, transpiration rate and nitrogen content in a tropical dry forest. *Physiologia Plantarum* **90**, 210–215.
Stearns, S.C. (1976). Life history tactics: A review of the ideas. *Quarterly Review of Biology* **51**, 3–47.
Stearns, S.C. (1989). Trade-offs in life-history evolution. *Functional Ecology* **3**, 259–268.
Steven, M.D., Biscoe, P.V. and Jaggard, K.W. (1983). Estimation of sugar beet productivity from reflection in the red and infrared spectral bands. *International Journal of Remote Sensing* **4**, 325–334.

Tans, P.P., Fung, I.Y. and Takahashi, T. (1990). Observational constraints on the global atmospheric CO_2 budget. *Science* **247**, 1431–1438.

Tilman, D. (1988). *Plant Strategies and the Dynamics and Structure of Plant Communities*. Princeton University Press, Princeton, NJ.

Tilman, D. (1991). Relative growth rates and plant allocation patterns. *American Naturalist* **138**, 1269–1275.

Tucker, C.J., Holben, B.N., Elgin, J.H. and McMurtry, J.E. (1981). Remote sensing of total dry matter accumulation in winter wheat. *Remote Sensing of Environment* **11**, 171–189.

Tucker, C.J., Vanpraet, C.L., Boerwinkel, E. and Gaston, A. (1983). Satellite remote sensing of total dry matter accumulation in the Senegalese Sahel. *Remote Sensing of Environment* **13**, 461–474.

Tucker, C.J., Fung, I.Y., Keeling, C.D. and Gammon, R.H. (1986). Relationship between atmospheric CO_2 variations and a satellite-derived vegetation index. *Nature* **319**, 195–199.

van der Werf, A., Visser, A.J., Schieving, F. and Lambers, H. (1993). Evidence for optimal partitioning of biomass and nitrogen availabilities for a fast- and slow-growing species. *Functional Ecology* **7**, 63–74.

Verma, S.B., Sellers, P.J., Walthall, C.L., Hall, F.G., Kim, J. and Goetz, S.J. (1993). Photosynthesis and stomatal conductance related to reflectance on the canopy scale. *Remote Sensing of Environment* **44**, 103–116.

Vitousek, P. (1982). Nutrient cycling and nutrient use efficiency. *American Naturalist* **119**, 553–572.

Walters, M.B. and Field, C.B. (1987). Photosynthetic light acclimation in two rain-forest Piper species with different ecological amplitudes. *Oecologia* **72**, 449–456.

Waring, R.H., Law, B.E., Goulden, M.L., Bassow, S.L., McCreight, R.W., Wofsy, S.C. and Bazzaz, F.A. (1995). Scaling gross ecosystem production at Harvard Forest with remote sensing: a comparison of estimates from a constrained quantum-use efficiency model and eddy correlation. *Plant, Cell and Environment* **18**, 1201–1213.

Waring, R.H., Landsberg, J.J. and Williams, M. (1998). Net primary production of forests: a constant fraction of gross primary production? *Tree Physiology* **18**, 129–134.

Warnant, P., François, L., Strivay, D. and Gérard, J.-C. (1994). CARAIB: A global model of terrestrial biological productivity. *Global Biogeochemical Cycles* **8**, 255–270.

Williams, K., Field, C.B. and Mooney, H.A. (1989). Relationships among leaf construction cost, leaf longevity, and light environment in rain-forest plants of the genus *Piper*. *American Naturalist* **133**, 198–211.

Yin, X. (1994). Nitrogen use efficiency in relation to forest type, N expenditure and climatic gradients in North America. *Canadian Journal of Forest Research* **24**, 533–541.

Generalist Predators, Interaction Strength and Food-web Stability

G.P. CLOSS, S.R. BALCOMBE AND M.J. SHIRLEY

I. SUMMARY

The recent description of complex food webs that contain many species, abundant omnivores, relatively long food chains and have a high number of links per species has cast significant doubt over our understanding of the relationships between food-web stability, interaction strength and overall

ADVANCES IN ECOLOGICAL RESEARCH VOL. 28
ISBN 0-12-013928-6

web complexity. Complex food webs contradict previous theoretical observations that highly complex webs should also be unstable and hence less likely to persist.

The description of a high links per species ratio in food webs suggests that generalist predators are a common feature of many systems. Despite the abundance of such predators, there have been few considerations of their role in food-web dynamics. Theoretical and empirical studies of generalist predators suggest that their role in food-web dynamics may range from being highly destabilising to having little impact on food-web dynamics. We argue that a full understanding of the role of generalist predators in food-web dynamics can only be developed when their actions are considered within a broad spatial and temporal framework.

In contrast to specialist predators, generalist predators have the capacity to switch to alternative prey as preferred prey decline. Consequently, the predator's population dynamics are not necessarily closely coupled to that of their preferred prey. When, considered over short temporal scales and at small spatial scales, such a predator is clearly destabilising as local extinction of vulnerable prey may occur given that the predator's abundance can be sustained by access to alternative prey. Species loss from a community is, however, a phenomena that is dependent on the spatial and temporal scales at which it is viewed. It cannot continue indefinitely and, once predators have eliminated vulnerable prey, those that remain are likely to be little affected by that predator's actions even though significant numbers of prey individuals may still be consumed by that predator. Remaining prey may persist within the predator's habitat, through protection within refugia or alternatively by maintaining reproduction rates that exceed the rate of predation. The food web persists in a stable, albeit altered form, but is now dominated by weak, effectively donor-control, interactions.

We review evidence for this perspective of generalist predation and food-web dynamics. Consideration of the role of generalist predators in food webs suggests that dynamic constraints on food-web structure may only apply in food-webs with few species, that the relative importance of top-down and bottom-up regulation of food-web structure may be strongly linked to the availability of refugia within a habitat, and that the strength and energetic importance of a link may, to some degree, be inversely related.

II. INTRODUCTION

The relationship between food-web stability, interaction strength and complexity has been a recurrent theme in ecological research (Elton, 1927; MacArthur, 1955; May, 1972; Pimm, 1982; Lawton, 1989; Schoener, 1989; Pimm *et al.*, 1991; Polis, 1991; Paine, 1992; Hall and Raffaelli, 1993; Polis and Strong, 1996). Early heuristic models argued that community stability increased with complexity given that a predator feeding on many prey would suffer less than a predator feeding on a single prey species should any one prey species become extinct (Elton, 1927; MacArthur, 1955). This view was challenged by May (1972, 1973) who observed that stability declined as web size or connectance increased in randomly connected food-web models. The incorporation of patterns observed in collections of real food-webs into dynamic food-web models based on Lotka–Volterra (L–V) equations suggested that overall food-web structure may play a role in stabilising complex webs (Table 1; May, 1973; Pimm, 1982; Lawton, 1989; Sugihara *et al.*, 1989; Cohen *et al.*, 1990). Such patterns include short food chains, a low and constant links per species ratio irrespective of the number of species in the web, and a relative absence of omnivory. Pimm's (1982) observation that strong predator–prey interactions are common in food-webs also suggested solid empirical support for theoretical basis of L–V models (see Pimm, 1982; Lawton and Warren, 1988; Lawton, 1989).

The issue of the distribution of interaction strength within a food-web is of crucial importance given that the stability properties of dynamic food-web models are highly sensitive to changes in interaction strength (May, 1973; DeAngelis and Waterhouse, 1987; Murdoch and Bence, 1987; Crawley, 1992). The observation that stability declines with increasing complexity (May, 1972) only holds if it is assumed that average interaction strength with-

Table 1
Mean values of food web statistics derived from analyses of food-web collections and studies of detailed food webs

Study	Mean No. of Species	Links/ Species	Mean Food Chain	Longest Food Chain
Food web collections				
Sugihara *et al.* (1989)	21	—	2.07	2.8
Cohen *et al.* (1990)	17	1.88	2.88	4.2
Schoenly *et al.* (1991)	24	2.20	2.89	7
Havens (1992)	38	3.89	—	—
Detailed food web studies				
Sprules and Bowerman (1988)	12	3.0	3.0*	—
Warren (1989)	22	5.0	3.8	5.9
Hall and Raffaelli (1991)	92	4.4	5 *	9.0
Martinez (1991)	182	13.0	7	15.0
Polis (1991)	30	9.6	7.3	12.0
Closs and Lake (1994)	40	2.0	1.8	2.4
Tavares-Cromar and Williams (1996)	34	3.2	—	—

*Indicates modal rather than mean food-chain length.
— Indicates statistic not calculated in particular study.

in a food-web is high. In contrast, a model which assumes that average interaction strength is zero, i.e. the donor-control (D-C) model (DeAngelis, 1975) exhibits a limited or no relationship between web complexity and stability (Figure 1). D-C-generated food-webs remain stable irrespective of features such as the number of links per species or degree of omnivory within the web (Lawton, 1989). The D-C model has, however, received little further attention given that its assumption of uniformly zero interaction strength and a predicted absence of stability related constraints on food-web structure did not correspond with the relatively frequent observation of strong predator–prey interactions in many communities, and the properties of food-webs observed in early meta-analyses of published food-web data (Table 1; Pimm, 1982; Lawton, 1989).

It is important to recognise that existing L-V (uniformly strong interactions) and D-C (uniformly zero interaction strength) food-web models only examine food-web dynamics at either end of the strong/weak interaction-strength continuum that is likely to occur even within a single food-web (Figure 1; Lawton, 1989; Abrams, 1993; Menge *et al.*, 1996; Hurlbert, 1997). Debate as to what is a realistic distribution of interaction strength across a food-web has continued and, contrary to earlier views (see Pimm, 1982), recent analyses of interaction strength in food-webs suggest that many, if not most, interactions in community food-webs are dynamically weak (Hall and Raffaelli, 1991; Polis, 1991, 1994; Paine, 1992; Raffaelli and Hall, 1992, 1996; Closs, 1996; Polis and Strong, 1996; Tavares-Cromar and Williams, 1996). In addition, analyses of the structure of food-webs collected specifi-

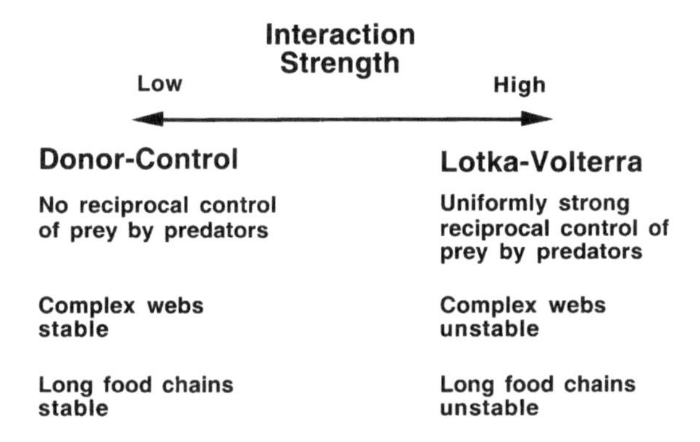

Fig. 1. Donor-control and existing Lotka–Volterra dynamic food-web models represent opposite ends of the continuum of interaction strengths that may occur within a food web (i.e. they assume interaction strength is either uniformly zero or strong respectively). The stability and structure of model food webs is dramatically affected by the assumed average interaction strength.

cally for the purpose of examining food-web structure also suggest that food-webs are far more complex than previously thought (Table 1; Sprules and Bowerman, 1988; Warren, 1989; Winemillar, 1989, 1990; Martinez, 1991, 1992; Polis, 1991; Havens, 1992; Closs and Lake, 1994; Tavares-Cromar and Williams, 1996). Such observations appear to contradict the assumptions and predictions of the L-V model directly (Lawton, 1995) and would appear to provide support for the D-C dynamic model. The D-C model, however, still cannot account for the occurrence of strong predator–prey interactions in real food-webs (Pimm, 1982). Whilst strong links may be relatively few in number, they can play a key role in determining food-web structure (Paine, 1980). Such observations suggest that there is a clear need to develop models that better reflect the distribution of linkage strength within food webs (Lawton, 1995).

The increasing dichotomy between the theoretical basis and predictions of existing dynamic food-web models, and empirical observation has prompted calls to re-examine food-web theory (Winemillar, 1989, 1990; Martinez, 1992; Paine, 1992; Abrams, 1993; Polis, 1994; Lawton, 1995). Despite the apparent failure of existing theoretical models, the challenge to develop new theory to explain and predict food-web structure and dynamics remains, given that May's (1972) theoretical observation that increasing food-web complexity results in reduced stability remains essentially unresolved. Complex food webs that contain some strong predator–prey interactions are being described, yet no current models of food-web structure and dynamics can account for their existence (Table 1; Lawton, 1995; Morin and Lawler, 1996).

We believe that a key element missing from our understanding of food-web dynamics is an appreciation of the role of generalist predators in a spatial and temporal context. Numerous dietary studies of individual predators, and the observation of abundant omnivory and high links per species ratios in recent analyses of detailed food webs (Table 1; Winemillar, 1990; Polis, 1991; Martinez, 1991, 1992; Closs and Lake, 1994; Lancaster and Robertson, 1995) suggest that generalist predators are a common feature of real webs. Generalist predators require special consideration given that their dynamics and impact can be expected to be distinctly different compared with that of specialised predators (see Andersson and Erlinge, 1977; Erlinge, 1987; Murdoch and Bence, 1987; Holt and Lawton, 1994; Reid *et al.*, 1997). Rosenzweig and MacArthur (1963) argued that predators could promote stability where there was access to alternative prey or where their population was limited by factors other than the availability of prey. In this paper we revisit and extend Rosenzweig and MacArthur's (1963) predictions and argue that an understanding of the impact of generalist predation in a spatial and temporal context can offer a unifying theme around which several aspects of food-web dynamics can be organised.

III. KEY DEFINITIONS: FOOD-WEB STABILITY AND INTERACTION STRENGTH

A. Food-web Stability

The concept of food-web stability has multiple ecological meanings, and potentially occurs at multiple levels of spatial and temporal scale (Connell and Slatyer, 1977; Hastings, 1988). In the following discussion, we consider mechanisms that might permit the coexistence, or alternatively drive extinction, of predator and prey populations at a local scale. In the context of the following discussion, this means at the scale at which the populations of a predator and its prey interact directly, i.e. no metapopulation dynamics. We believe the question is important, given that the simplest mathematical models of predator–prey dynamics predict a low probability of stability and persistence amongst directly interacting predator and prey populations (Murdoch and Oaten, 1975). As highlighted by May (1972, 1973), the instability of simple systems poses a real challenge for ecologists faced with the challenge of explaining how complex systems actually persist.

In a mathematical context, a stable system is one in which species' populations return to a pattern of predictable behaviour following perturbation (May, 1973; Murdoch and Oaten, 1975; Pimm, 1982; DeAngelis and Waterhouse, 1987; Lawler, 1993). However, in empirical systems it is difficult to use such a measure to ascertain system stability (Pimm, 1982; Lawler, 1993). More tractable empirical parameters that may be correlated with mathematical stability include population variability and population persistence (Pimm, 1982; Lawler, 1993). Population variability and persistence have been used in previous comparisons between the stability of real food webs and the predictions of mathematical models (see Pimm, 1982). Mathematical models that are unstable show greater variability in abundances compared with those that are stable, and there is empirical evidence which suggests that highly variable populations are more likely to become extinct compared with less-variable populations (Pimm, 1982; Lawler, 1993). Variation in abundance is, however, a natural feature of the life cycle of many organisms, particularly for those species which might be described as opportunist, early colonising species (Connell and Slatyer, 1977). Such temporal variation in abundance may occur irrespective of the dynamics of predator and prey within the system (Connell and Slatyer, 1977). In the context of this paper, we define a stable system as one where persistence of predator and prey occurs at a local scale indefinitely. A less-stable system is one where predators drive vulnerable prey locally extinct.

B. Interaction Strength

The concept of predator–prey interaction strength is central to the food-web stability–complexity debate, yet there is considerable uncertainty as to what it is and how to measure it (Lawton, 1989; Goldwasser and Roughgarden, 1993; Menge *et al.*, 1996; Raffaelli and Hall, 1996; Tavares-Cromar and Williams, 1996; Benke and Wallace, 1997; Hurlbert, 1997). MacArthur (1972) originally defined predator–prey interactions as either weak or strong. Strong interactors are predators or consumers whose removal results in a subsequent substantial change in community structure. Alternatively, little or no change results if a weak interactor is removed (MacArthur, 1972). Interaction strength thus defined can only be determined through manipulative predator removal–introduction experiments (Paine, 1980, 1992). Statistics that provide some measure of energetic flux across particular trophic links have also been used as a measure of interaction strength (e.g. Goldwasser and Roughgarden, 1993; Tavares-Cromar and Williams, 1996; Benke and Wallace, 1997). Such measures may involve simple estimates of energy flux based on the frequency of a particular prey item in the diet of a predator (e.g. Goldwasser and Roughgarden, 1993; Tavares-Cromar and Williams, 1996), or considerably more sophisticated estimates of per capita prey production and consumption by predators (e.g. Benke and Wallace, 1997).

It is important to note that the measurement of interaction strength by either the removal or addition of a predator, or by the determination of the per capita impact of predators on prey populations (e.g. Paine, 1992) both represent a very different measure of food-web structure compared with measurements based on the determination of energetic flux across trophic links (e.g. Goldwasser and Roughgarden, 1993). As we discuss later, attempts to use such measures interchangeably (e.g. Goldwasser and Roughgarden, 1993; Tavares-Cromar and Williams, 1996) are inappropriate (see also Polis and Strong, 1996). In the context of the dynamics of L–V and D-C food-web models, the only appropriate empirical measure of interaction strength is the degree of reciprocal regulation of predator and prey population dynamics (see Pimm, 1982). In the following discussion, we shall use the term interaction strength as originally envisaged by MacArthur (1972), i.e. referring to the reciprocal degree of influence that predators and prey may have on each others' population dynamics. A strong L–V interaction is one where predators and prey exert strong reciprocal regulation over each others' population dynamics (Pimm, 1982). In contrast, a D-C interaction assumes that predators exert no control over the population dynamics or abundance of their prey, although the abundance of prey may influence the population dynamics of the predator (DeAngelis, 1975). Significantly, such a definition does not make any reference to the energetic importance of any particular link. As we shall argue subsequently, any particular L–V or D-C interaction may support high or low levels of energy flux.

IV. POPULATION DYNAMICS OF GENERALIST PREDATORS

Few studies have specifically considered the impact of generalist predation on food-web dynamics (Murdoch and Bence, 1987; Lawton, 1989). Generalist predators present a degree of challenge to classical dynamic equilibrium models of predator–prey systems given that their dynamics are not strongly coupled to the abundance of any one prey species (Murdoch and Bence, 1987; Lawton, 1989; Crawley, 1992). Empirical evidence suggests that the impact of generalist predators may vary greatly in space and time with various authors observing that generalist predators may: (1) destabilise given that they cause local extinctions (Murdoch and Bence, 1987); (2) stabilise communities by suppressing dominant species and restricting prey population eruptions (Rosenzweig and MacArthur, 1963; Paine, 1966, 1980; Reid *et al.*, 1997); and (3) play little active role in the regulation of communities given that the impact of their predation is dispersed across many prey species, although they may have previously determined species composition when first introduced to a particular habitat (Thorp, 1986; Lawton, 1989; Strong, 1992).

The simplest generalist predator–prey mathematical models assume that predator and prey population dynamics are uncoupled, i.e. they have no influence on each others' abundance (Crawley, 1992). In simple uncoupled models of a generalist predator and its prey, any equilibria that occur tend to be trivial. Perturbations to the system tend to result in either runaway prey population growth or prey extinction (Crawley, 1992). In a theoretical context, such instability results in system collapse and is hence considered unrealistic (Crawley, 1992). Models of uncoupled predator–prey systems can be stabilised by including either refuges for prey or assuming a predator functional response that results in reduced predation rates as prey abundance declines (Crawley, 1992).

Such simple models are clearly deficient in a number of respects. Simple uncoupled generalist predator–prey models such as those described by Crawley (1992) assume that generalist predators feed on a uniform prey population, i.e. all species are equally vulnerable. However, generalist predators are rarely, if ever, complete generalists. They invariably show preferences for certain prey within their habitat. Selectivity for prey may be influenced by a myriad of factors such as prey size, colour, predator avoidance abilities, abundance (see Townsend *et al.*, 1986; Sih, 1987; Lima and Dill, 1990; Abrams, 1993). Certain preferred species will suffer a disproportionate level of predation pressure (Murdoch and Bence, 1987). Significantly, the capacity of generalist predators to feed on multiple prey species can allow them to depress populations of vulnerable, preferred prey to very low levels or even drive them extinct (at least at a local scale), whilst maintaining their own abundance by preying on less vulnerable, alternative prey (Murdoch and Bence, 1987). Such a situation would appear to be analogous to the "apparent competition" model proposed by Holt (1977).

A further deficiency with many existing predator–prey models is that all assume that predators are limited solely by prey/food availability. It is well known, however, that many factors other than food may limit a predator's abundance (Rosenzweig and MacArthur, 1963; Polis and Strong, 1996). In streams, the density of adult brown trout, *Salmo trutta* L. may be determined by the availability of suitable feeding sites for adults (Bachman, 1984) or, alternatively, by density-dependent factors operating during early life-history stages (Elliot, 1994). The abundances of adult yellow perch, *Perca flavescens* Mitchell, and European perch, *P. fluviatilis* L., are frequently determined by the impact of various environmental factors or predation and competition on juvenile stages, rather than the availability of food to adult perch (Persson, 1987; Persson and Greenberg, 1990a,b; Hartman and Margraf, 1993). Numerous similar examples can be found in terrestrial systems, e.g. the availability of territories for common buzzards, *Buteo buteo* L., and foxes, *Vulpes vulpes* L. (Erlinge, 1987). Many generalist predators are also relatively long-lived and breed seasonally, whilst their prey are short-lived and may breed several times over the one breeding season, e.g. fish–zooplankton systems (Polis and Strong, 1996). This has the potential to severely constrain the capacity of predators to respond numerically to increasing prey abundance over the short term in the absence of any immigration. Should prey abundance exceed the predator's capacity to respond numerically to increasing prey abundance, prey populations will escape predator control (Pech *et al.*, 1992; Begon *et al.*, 1996a; Polis and Strong, 1996).

The difficulties associated with modelling stable, realistic predator–prey systems has led to an increasing level of doubt as to whether stability is a property of predator–prey systems, at least at local spatial and temporal scales (Murdoch *et al.*, 1985; DeAngelis and Waterhouse, 1987; Murdoch, 1994). Indeed, at the level of the individual predator and prey, all predator–prey interactions involve extinction (Murdoch *et al.*, 1985; DeAngelis and Waterhouse, 1987). However, extinction at a local patch scale may occur regularly, particularly if predators are generalists and their dynamics are uncoupled to prey dynamics at a patch level (Murdoch *et al.*, 1985; Murdoch and Bence, 1987; Murdoch, 1994). Coexistence of predator and prey, and hence the stability of the system at larger scales, can be maintained by either recolonisation of patches by prey from nearby predator-free areas, i.e. metapopulation dynamics (Hastings, 1978), through the protection of proportions of prey populations inside refuges (Sih, 1987; DeAngelis and Waterhouse, 1987; Murdoch and Bence, 1987), or prey switching by predators (Townsend *et al.*, 1986). These models are significant in a food-web context given that they suggest that interaction strength (as defined by MacArthur, 1972) may vary in space and time.

V. PUTTING GENERALIST PREDATORS INTO FOOD WEBS: A PERSPECTIVE THAT INCORPORATES SPATIAL AND TEMPORAL SCALE

A. Generalist Predators in Spatially Simple Systems

The impact of a generalist predator on community dynamics in space and time can be most clearly seen by following the changes that occur within a community after a predator is introduced to a new system. Consider a system of many potential prey species to which a small population of a generalist predator is then introduced (Figure 2). All species within the community can in theory be considered to be potential prey. Potential prey species within the community will vary from having only limited or no vulnerability to the predator (e.g. too small to be attractive) to total vulnerability (e.g. no capacity to escape and highly visible). Prey species also vary with respect to their capacity for population increase (Hulsmann and Mehner, 1997). Therefore,

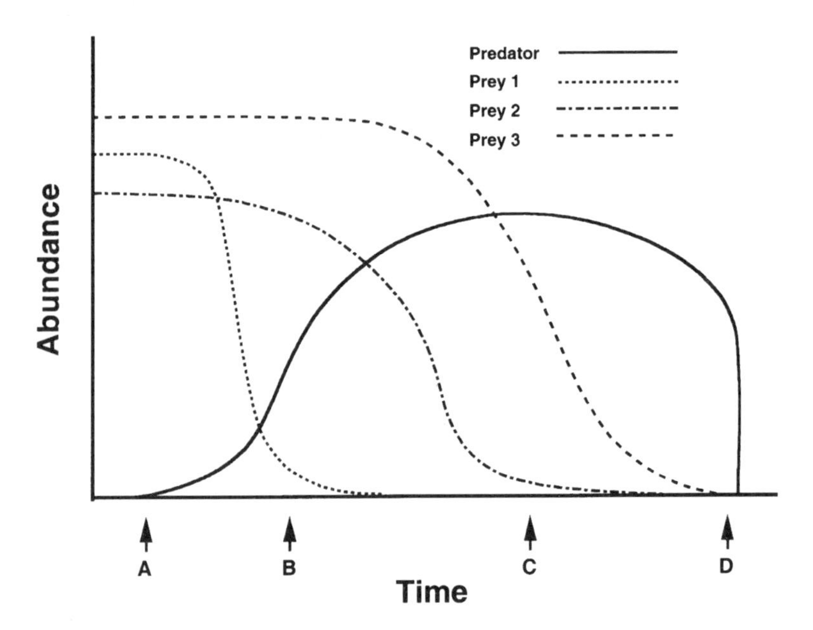

Fig. 2. Patterns of predator and prey abundances predicted to occur through time following the introduction of a predator to a spatially restricted, structurally simple habitat. Point A represents the time of predator introduction. At point B, prey 1 is virtually eliminated and predatory pressure on prey 2 increases as the predator increasingly switches to the alternative prey. Points C and D identify the near-elimination or extinction of successive prey 2 and 3, owing to increasing predator consumption as predator abundance continues to increase. Shortly after point D, predator extinction also occurs due to the elimination of all available prey.

the capacity of any prey species to maintain a population under predation pressure will be determined by a combination of their vulnerability to the predator and the capacity of that particular prey species for population increase (Hulsmann and Mehner, 1997).

If abundant prey are available to the predator and no other factors are limiting, the population of the predator will grow, resulting in a progressive increase in the average intensity of predation across the community (Begon *et al.*, 1996a). Populations of different prey species will persist as long as the capacity of any particular prey species for population increase exceeds the rate of mortality due to predation. If the population of the predator continues to increase, the rate of mortality due to predation will ultimately begin to exceed the capacity of certain prey species to reproduce themselves. Such species will inevitably become extinct or at least decline to very low abundances (Figure 2). Crucially, in the case of a generalist predator, extinction of a single prey species will not necessarily limit the predator population should the predator have access to, and be capable of, switching to alternative prey (Crawley, 1992), although the growth of the predator population may be reduced given the loss of access to the most preferred prey. Assuming that the predator is constrained only by food, the predator could keep increasing as it progressively eliminates successive prey species and then switches to the next preferred prey species. Ultimately, a one predator–one prey system will result and the system will either crash as the predator eliminates the last available prey species (Figure 2) or a dynamic reciprocal equilibrium will be established (Figure 3). Such a model could be viewed as a multi-species extension of the "apparent competition" concept (Holt, 1977). Should the predator be removed, the process may reverse with vulnerable prey species recolonising the habitat, although such a process may take some time as prey species progressively locate the newly created "enemy-free-space" (see Macan, 1965, 1966, 1977; Jeffries and Lawton, 1984).

B. Generalist Predators in Spatially Complex Systems

The situation described above is perhaps analogous to that which has been described from structurally simple, spatially limited habitats, e.g. a fish–zooplankton system in a small pool (e.g. Hurlbert and Mulla, 1981) or simple protist predator–prey systems (Morin and Lawler, 1996). Complex, multi-species communities with many generalist predator species and prey do, however, exist and persist in nature (see Polis and Strong, 1996). The potentially stabilising nature of generalist predation only becomes apparent when its impact is considered in spatially larger and more complex systems over a period of time (Figure 4). Consider the situation that develops following the introduction of a predator into a large, spatially complex habitat containing a variety of potential prey species. Access to abundant prey again results in

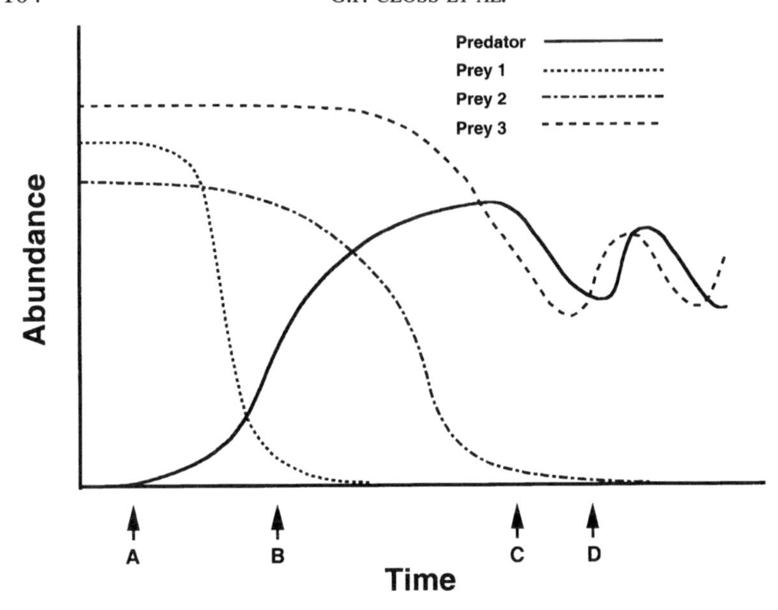

Fig. 3. Identical dynamics to those represented in Figure 2 until point D. At point D, the predator and prey 3 enter into a dynamic Lotka–Volterra equilibrium in which reciprocal control of predator and prey population dynamics occurs.

rapid predator population growth. Again, the likelihood of prey extinction will be determined by a combination of prey vulnerability and reproduction rate. As the predator population and hence predation rate increases, the predator will begin eliminating prey from its local environment. However, assuming that the predator's numerical response to increasing prey abundance will be ultimately constrained in some way, four categories of interaction strength can be envisaged. These are listed below.

1. Dynamic Strong Interactions

Those prey species which suffer a mortality rate greater than their reproduction rate and have no refuge within the habitat will ultimately become extinct within that habitat (Figure 4). The initial phase of the interaction will in fact behave as a dynamic reciprocal interaction given that predator numbers will be increasing in response to the high availability of prey, and prey numbers will begin to fall as predator pressure increases. Once the prey are eliminated from the predator's habitat, however, the interaction will cease to exist in any significant energetic sense. Prey which fall into this category may form a major component of the predator's diet during the period immediately after

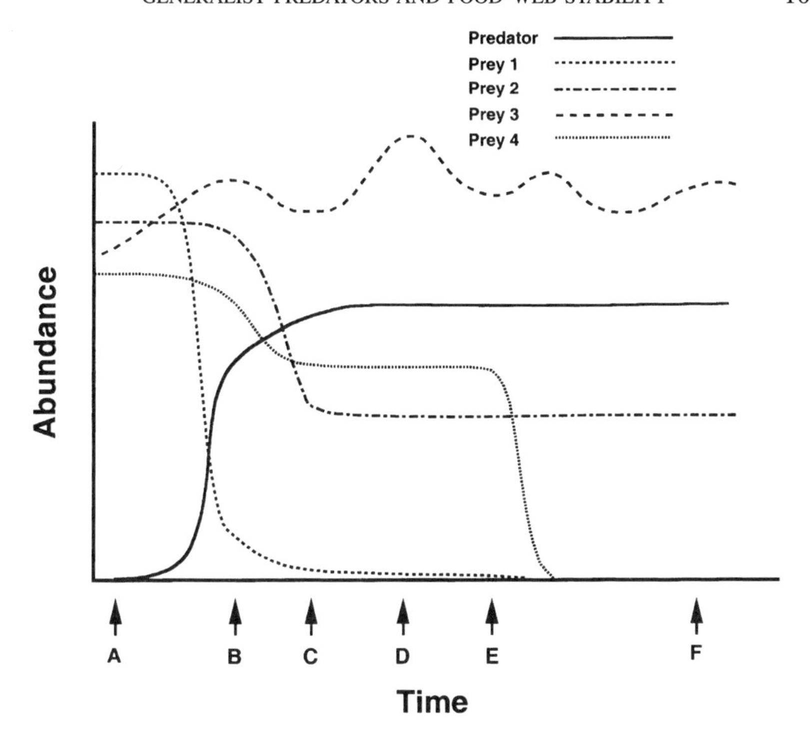

Fig. 4. Patterns of predator and prey abundance predicted to occur through time following the introduction of a predator to a spatially extensive and/or structurally complex habitat. Point A represents the time of predator introduction. At point B, prey 1 is virtually eliminated due to increasing predatory pressure, resulting in a switch in predator feeding to alternative prey. This point represents the effective end of dynamic strong interactions occurring within the habitat. Prey 2 and 4 begin to decline, but the predator has little or no impact on the population dynamics of prey 3. At point C, prey 2 is restricted to some form of spatial refugia, resulting in a strong non-dynamic interaction, but the predator continues to interact only weakly with prey 3. Prey 4 enters into a locally stable but globally unstable equilibrium with the predator. At point D, predator densities reach a constant level, owing to limits on population growth other than food or overall food limitation. At point E, some form of perturbation reduces the abundance of prey 4, upsetting the locally stable equilibrium and resulting in its ultimate elimination from the habitat. At point F, only two prey species persist indefinitely within the habitat. Prey 2 is restricted to a refuge with predators feeding on surplus prey production from that refuge. Reproduction rates of prey 3 exceed the rates of predation, hence the predator has little or no impact on population dynamics. All interactions from point F onwards are effectively donor-control interactions, i.e. the predator has no impact on prey population dynamics, although rates of prey production may determine predator abundance.

predator introduction; however, once eliminated, they will no longer occur in the diet given that they will not be locally available (e.g. Oscarson, 1987). An important consequence of this pattern is that the very strongest interactions between a particular generalist predator and its prey are likely to be the least important energetically over longer time scales (see Lawton, 1989; Strong, 1992).

2. Non-dynamic Strong Interactions

A second category of prey will be those species which become restricted to some form of refuge (Figure 4; see also Rosenzweig and MacArthur, 1963). Refuges from predation may come in many forms including size, spatial and temporal refuges (Jeffries and Lawton, 1984); however, a key feature is that they are either wholly or partially closed to the predator (Jeffries and Lawton, 1984; Crawley, 1992). Prey species which have a proportion of their population protected within a refuge may still, however, form a significant proportion of the predator's diet, given that the predator may feed on surplus production emerging from the refuge. Crucially, the predator does not exert reciprocal control over prey abundance given that prey production and abundance within the refuge is determined by the availability of resources within that refuge and not by the predator (see Errington, 1946; Mittelbach, 1988). Predator production may still, however, be determined by surplus prey production emerging from the refuge, hence the interaction is, in effect, a donor-control interaction (see DeAngelis, 1975).

3. Weak Interactions

If the predator is unable to respond numerically and functionally to prey species which are capable of high reproduction rates, a third category of prey which are beyond any effective control by the predator can exist (Figure 4). The inability of a predator to increase its numbers in response to high prey abundance may be due to permanent (e.g. limited breeding sites, juvenile bottlenecks, etc.) or temporary (a single breeding season per year) constraints. In either case, the effect is the same, i.e. no instantaneous reciprocal control over prey abundance. Prey in this category may again form an energetically significant proportion of a predator's diet. However, the interaction strength of such links will be negligible given that the predator is not capable of regulating prey production and abundance. Again, the interaction will exhibit donor-control dynamics (DeAngelis, 1975). Prey abundance will presumably be regulated by other density-dependent and independent factors not related to the particular generalist predator under consideration (see also Power, 1992).

4. Equilibrium Interactions

Prey may also exist in a dynamic equilibrium with predators should the rate of predation happen to equal the rate of prey production (Figure 4). Such an equilibrium will be locally unstable given that predator and prey dynamics are not closely coupled. Should prey production rate exceed the rate of predation, the prey population will cease to be regulated by the predator. Should the predation rate exceed the prey production rate, the prey species will become locally extinct. Divergences away from the local equilibrium could conceivably be driven by density-independent factors such as changes towards more or less favourable environmental conditions for the prey. Given that such dynamics will be globally unstable (Crawley, 1992), few interactions of this type are likely to be seen or persist should they occur.

C. Food Webs as Dynamic yet Persistent Entities

The preceding discussion has clear parallels with the analysis of Rosenzweig and MacArthur (1963) in that it argues that predators will switch to alternative prey when preferred prey become scarce, that prey may be protected within refugia and that predators may be limited by factors other than food. It does, however, go further in suggesting that particularly vulnerable prey species may be driven to local extinction and that many prey species within a food-web are unlikely to be regulated by predators. In the period immediately after the introduction of a generalist predator, dynamic strong interactions may be common, possibly representing a significant proportion of the energy flux through the food-web. Such links are likely to be transient as vulnerable prey species become extinct. In L–V models, the loss of species from a web is considered to represent instability. Local extinction due to predation does, however, occur in real food webs. Crucially, the process of species loss is transient in space and time. Once a predator has eliminated vulnerable species, a food-web will remain, albeit in an altered form. Thus, a food-web into which a generalist predator has been introduced is a persistent yet dynamic entity, with potentially conspicuous changes occurring through space and time. Once any vulnerable species are eliminated, dynamic strong interactions will cease to exist in any significant energetic sense, and non-dynamic strong and weak interactions tend to dominate. Such webs will tend to be dominated by donor-control dynamics, and hence any relationships between overall web complexity (i.e. connectance, food-chain length and omnivory) and stability are likely to be weak.

The dynamic nature of food webs is well illustrated by a process that occurs should the predator be removed. Species formerly restricted to refugia tend to expand out from refugia and previously excluded species recolonise (Macan, 1977; Closs, 1996; Closs and Lake, 1996). The exact pattern of change will presumably vary somewhat as chance events may influence the precise sequence in which previously excluded prey colonise the newly available "enemy-free space". A period of community instability may result as a series of new and potentially intense predatory and competitive interactions work their way through the community as formerly vulnerable species begin to exclude species that were favoured under the previous predatory regime. The effects of such predator removals have been well studied in lentic systems where the removal of fish may trigger a cascade of community change as large predatory water bugs colonise and expose a range of zooplankton species that were capable of avoiding fish predation to a new predatory threat (Zaret, 1980).

VI. IMPLICATIONS FOR FOOD-WEB THEORY

A. Interaction Strength and its Relationship with Energy Flux

Consideration of the dynamics of a generalist predator in space and time suggests the potential for some degree of relationship between linkage strength and the energetic flux across any particular trophic link. We concur with several previous authors who have either clearly stated or alluded to the observation that the strongest interactions also tend to be the least important energetically (Paine, 1980, 1992; Thorp, 1986; Lawton, 1989; Strong, 1992; Morin and Lawler, 1996; Polis and Strong, 1996). Numerous studies in a variety of habitats have demonstrated that prey species most vulnerable to a particular predator tend to be totally excluded from the habitats occupied by that predator and hence are no longer available as prey (see Figure 5: sea urchin, *Strongylocentrotus purpuratus*–benthic diatom link; Paine, 1980) (see also Zaret, 1980; Fairweather, 1987; Oscarson, 1987).

The observation that strong links are, in an energetic sense, the least important highlights an important point. It suggests that, once vulnerable prey are eliminated, a generalist predator will tend to exert little or no reciprocal control over the remaining prey species upon which it feeds. The argument that interaction strength and energy flux are inversely related across any particular trophic link is, however, somewhat simplistic given that it does ignore the possibility that strong non-dynamic interactions can indeed be energetically significant (see Figure 5: kelp, *Hedophyllum*–chiton, *Katharina* link; Paine, 1980). Our model of generalist predation suggests a clear mechanism for the persistence of trophic links characterised by high interaction strength and high energetic flux. Numerous studies have demonstrated the restriction of vulnerable species to spatial refuges such as areas of aquatic macrophytes or complex substrates, with predators preying upon surplus production from that refuge (e.g. Macan, 1965, 1966; Dayton, 1971; Menge and Lubchenco, 1981; Gilinsky, 1984; Menge *et al.*, 1985; Oscarson, 1987; Diehl, 1995).

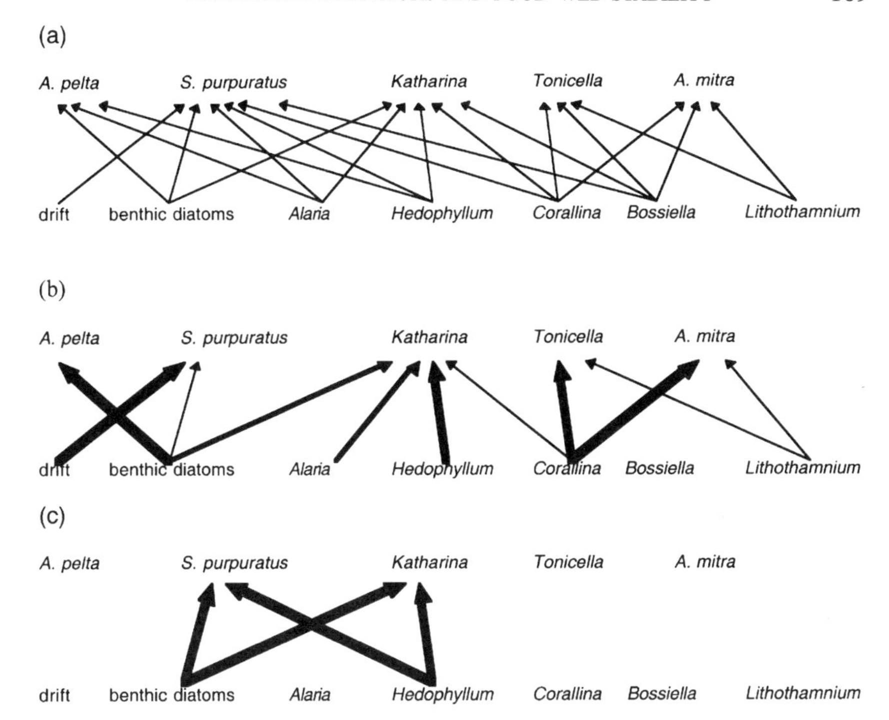

Fig. 5. Three conceptually and historically different approaches to depicting trophic relationships, illustrated for the same set of species. The connectedness web (a) is based on observation, the energy flow web (b) on some measurement and literature values, and the functional web (c) on controlled manipulation. Relationships between the strength of and interaction, and the energetic flux across a particular link may vary greatly. (Reproduced from Paine (1980), with permission from R.T. Paine and Blackwell Scientific Publications.)

The proportions of weak, non-dynamic strong and dynamic strong interactions within a habitat will also vary with respect to space and time. Predators which have been present in a particular habitat for a long period of time may or may not be food limited. If the predator is food limited, average prey production from refuges will be insufficient to maintain maximum predator production. Predators may exhibit reduced feeding or growth rates even though prey remain abundant within any available refugia (Crowder and Cooper, 1982; Diehl and Eklov, 1995). Predatory pressure may, however, still be sufficient to eliminate particularly vulnerable prey species (Gilinsky, 1984). In a community where a generalist predator is food limited, a prey community dominated by prey species that are limited to refugia would appear likely, hence strong non-dynamic interactions will dominate (Figure 6). High rates of energy flux across the strong non-dynamic links within the community may occur, although not necessarily so.

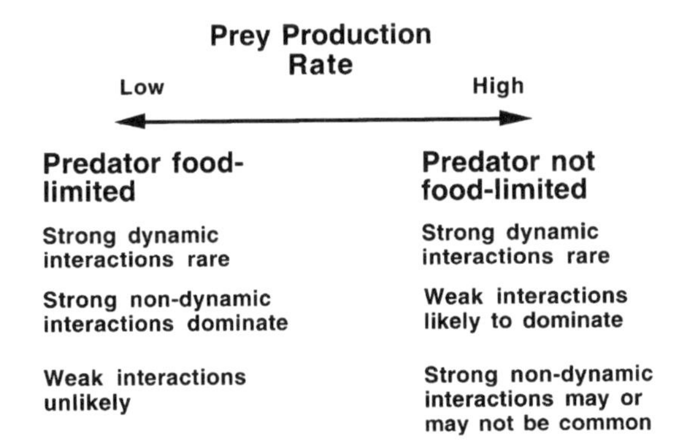

Fig. 6. The proportions of the four classes of interaction may vary with the production rate of prey. Strong non-dynamic interactions are likely to dominate in situations where predators are food-limited. In situations where predator abundance is limited by factors other than food, either weak or strong non-dynamic interactions may dominate.

In contrast, if surplus prey production from refuges is high, predators are likely to be limited by factors other than food (Figure 6). Predators may still drive particularly vulnerable prey extinct but those prey species with high breeding and recruitment rates can potentially escape predator-regulatory control. Consequently, weak interactions may be common within such food webs. High rates of energy flux may occur across these links, with predators consuming large numbers of such prey but having no significant impact on their population dynamics. Spectacular eruptions of prey species and community switches to alternative stable-states are likely to occur in such communities, particularly where the productivity of prey has the potential to be high, given conditions conducive to successful breeding and recruitment (see May, 1977; Pech *et al.*, 1992).

B. Relative Importance of Top-down versus Bottom-up Control within Food Webs

The debate regarding the relative importance of top-down and bottom-up regulation within food webs has been vigorous in recent years with many examples of either strong top-down or bottom-up regulation being demonstrated (see reviews by Hunter and Price, 1992; Menge, 1992; Power, 1992; Strong, 1992; Rosemond *et al.*, 1993; Oksanen *et al.*, 1996; Persson *et al.*, 1996). Various opinions as to the relative importance of top-down or bottom-up forces have emerged including the possibility of simultaneous top-down and bottom-up control (Hunter and Price, 1992; Strong, 1992; Rosemond *et al.*, 1993), the likelihood that top-down regulation will be of lesser importance in complex, speciose systems compared with simple, less speciose systems (Strong, 1992), and the dominance of bottom-up over top-down forces (Power, 1992). In our model of generalist predation, simultaneous top-down and bottom-up control will occur in most food webs, with top-down regulation determining prey distribution and bottom-up regulation determining secondary production (see also Power, 1992).

Strong top-down regulation of food-web structure undoubtedly occurs, having been described from a variety of aquatic and terrestrial systems (see reviews by Power, 1992; Strong, 1992; Menge *et al.*, 1994). Typically, evidence for strong top-down regulation occurs following changes in the abundance of a top predator (introduction or removal), a pattern consistent with our model of generalist predation. Generalist predators tend to either determine species composition or species distribution within a habitat (Thorp, 1986; Strong, 1992; Closs, 1996). In the period immediately following the introduction of a predator, strong top-down forces are likely to dominate. However, once species composition and distribution has stabilised, bottom-up forces will tend to gain primacy. Strong top-down influences remain, given that the predator will continue to determine prey distribution; however, this does not involve active regulation of prey-population dynamics within the refugia. In the longer term, and assuming that predators are unable to make rapid numerical responses to changes in prey abundance (the situation in most relatively short-term predator manipulations to date), predators will either feed on surplus prey production from refugia or on prey populations over which they exert little dynamic reciprocal control (see Power, 1992). In these food webs, prey production will be determined by the availability of resources rather than the impact of predation, i.e. bottom-up control. Dynamics of this nature have been described from many systems (see Thorp, 1986; Power, 1992; Menge *et al.*, 1994).

The importance of top-down or bottom-up regulation is also likely to vary with respect to the number of species in a web. In webs with few species, the number of food chains will inevitably be limited. Changes in the relative abundances of species within any one chain in a simple community will inevitably have significance at a community level simply because any one species will represent a significant proportion of that community. In more speciose webs containing multiple top to bottom food chains, any one predator is likely to interact with numerous prey species (see Polis, 1991), and will have varying impacts on the population dynamics of the species with which it interacts. Interactions between a generalist predator and its prey are likely to range through dynamic strong, non-dynamic strong to weak interactions, although it is the latter two interactions that are likely to dominate over longer periods of time. This pattern of either dynamic or non-dynamic strong interactions embedded amongst a web of weak interactions has been described from numerous food webs in both aquatic and terrestrial habitats (see Figure 5; Paine, 1980; Strong, 1992).

Stream food webs, in particular, may be typical of complex, speciose food webs in which a few strong predator–prey interactions appear embedded amongst numerous weak interactions (e.g. Power, 1990; Bechara *et al.*, 1992, 1993; Flecker and Townsend, 1994; Closs, 1996; Dudgeon, 1996). Stream communities are typically dominated by generalist fish or invertebrate predators (Allan, 1995), and benthic stream invertebrates are likely to be protected to a significant extent from fish predation by the complex substrate in which they live (Ware, 1972; Brusven and Rose, 1981; Flecker and Townsend, 1994; McIntosh and Townsend, 1994, 1995). In addition, many prey species appear to utilise flexible predator-avoidance strategies (*sensu* Sih, 1987) as a means of avoiding predators (e.g. McIntosh and Townsend, 1994; McIntosh and Peckarsky, 1996), thus further weakening the impact of predators on their potential prey.

C. The Relationship between Food-web Complexity and Stability

An increasing number of studies and reviews have suggested that dynamically weak links do in fact dominate many food webs (e.g. Thorp, 1986; Lawton, 1989; Paine, 1992; Closs, 1996; Morin and Lawler, 1996; Polis and Strong, 1996), an observation that would appear to directly contradict Pimm's (1982) assertion that an abundance of strong interactions place tight constraints on food-web structure. Pimm's observation that strong dynamic links are common within food webs rests on the observation that changes in species composition occur following predator manipulation. The period immediately following the addition or removal of a generalist predator, however, would be just when we would expect dynamic strong interactions and changes in species composition to be most evident.

Our model of generalist predation suggests that strong dynamic reciprocal interactions between generalist predators and their prey population are transient, consequently dynamically weak interactions and non-dynamic strong interactions will tend to dominate over longer temporal scales. Crucially, the stability of such a food-web will be a product of the stability of each predator–prey interaction within the web rather than an emergent property of overall food-web structure (see also Allen-Morley and Coleman, 1989). Such a pattern of dynamics is more akin to that predicted by D-C dynamics rather than L–V dynamics (DeAngelis, 1975; Pimm, 1982; Polis and Strong, 1996). A dominance of dynamically weak links within food webs suggests that there will also be a weak relationship between the overall web structure and stability, a pattern that has been suggested by experimental microcosm studies (Allen-Morley and Coleman, 1989) and recent analyses of detailed food webs (Warren, 1989; Winemillar, 1990; Polis, 1991; Martinez, 1992; Closs and Lake, 1994; Lancaster and Robertson, 1995; Closs, 1996; Tavares-Cromar and Williams, 1996). Indeed, our model suggests that if web stability is at all related to web complexity, it will be that stability is positively related to complexity given that a generalist predator with many prey is less likely to be adversely affected should one prey species go extinct compared with a predator with few prey, a line of argument first developed by Elton (1927) and extended by MacArthur (1955). If overall constraints on food-web structure exist, we suggest that they are more likely to be due to factors such as inefficiencies in energy transfer between trophic links, the spatial scale and complexity of the habitat, and the nature of the local species pool (see also Jenkins *et al.*, 1992; Spencer and Warren, 1996a,b).

A positive relationship between complexity and stability is supported by several lines of empirical evidence. Lawler and Morin (1993) observed that in microcosm protist food webs, populations of omnivorous predators tended to fluctuate less compared with specialist predators. Fagan (1997) also observed a similar pattern of increasing stability in arthropod assemblages containing omnivorous predators compared with assemblages containing only non-omnivorous predators. Studies of the population dynamics of small mammal species in northern Europe (Hansson, 1987; Reid *et al.*, 1997) suggest that population dynamics may be more stable in species-rich systems dominated by generalist predators compared with less speciose systems dominated by specialist predators. In harsh, high-latitude, low-diversity systems, specialist predators appear to drive cyclic variation in the abundance of small mammals (Hansson, 1987; Reid *et al.*, 1997). In contrast, small mammal diversity in lower latitude systems is comparatively high with greater production, thus allowing a range of generalist predators to survive (Hansson, 1987; Lindstrom, 1994). These generalist predators preferentially feed on the most abundant prey species, thus preventing prey eruptions and maintaining stable prey numbers (Hansson, 1987; Reid *et al.*, 1997). Unstable population dynamics are also a feature of the relatively simple food webs that develop in the pelagic zones of freshwater lentic systems (McCauley and Murdoch, 1987, 1990), whilst the complex species-rich webs that are found in benthic lentic and lotic systems appear to be characterised by a few strong interactions embedded amongst many weak interactions (Power, 1992; Flecker and Townsend, 1994; Closs, 1996).

The observation that simple webs may be more unstable suggests that dynamic constraints may influence overall food-web structure under certain circumstances. A number of studies have found a positive relationship between the number of species in a web and food-chain length (Hall and Raffaelli, 1991; Martinez, 1991; Martinez and Lawton, 1995; Bengtsson and Martinez, 1996; Spencer and Warren 1996a,b). In our model of generalist predation, any one predator–prey interaction can have significant implications for surrounding species within a web (Power, 1992; Strong, 1992). Consequently, in webs containing few species, a single influential species may affect the population dynamics of a significant number of species within the web and, hence, the overall structure of the food-web. In a web containing few species, it would appear likely that long food chains will be less stable given that the loss of any one prey species will tend to interrupt the flow of energy significantly to predators higher up (see Pimm, 1982; Fagan, 1997). In complex webs dominated by generalist predators and weak links, the loss of one link is unlikely to disrupt the entire web (Elton, 1927; MacArthur, 1955; DeAngelis, 1975; Fagan, 1997). If so, long food chains will be more likely to be observed in complex food webs, although the energetic flux along any particular food chain may well be trivial.

VII. FUTURE DIRECTIONS FOR FOOD-WEB RESEARCH

Our examination of the dynamics of generalist predators suggests that stability is not an emergent property of overall food-web structure but rather a property of the individual links that make up a food-web (see also Allen-Morley and Coleman, 1989). If so, questions aimed at examining food-web stability should focus on factors determining the stability of individual interactions and the dynamics of species interacting closely together within a food-web (e.g. food chains), rather than at the level of the entire food-web. Our emphasis on the importance of individual interactions within food webs also suggests that the practice of aggregating species into trophic groups (e.g. Spiller and Schoener, 1996; DeRuiter *et al.*, 1996) will tend to obscure the dynamics of individual predator–prey interactions within a web, in particular potential relationships between interaction strength and energetic flux. In the following section, we outline several research directions which focus at the scale of individual interactions and food chains rather than at the level of an entire food-web. Of particular interest are relationships between interaction strength and energetic flux, relationships between prey survival in one generation and prey abundance in subsequent generations, factors influencing the relative proportions of weak, non-dynamic strong and dynamic strong interactions within a food-web, the role of refugia in promoting stability and influencing the production of prey, and the development of theoretical models that model the dynamics of generalist predators within food webs.

A. Relationships between Interaction Strength and Energetic Flux

Unravelling the relationship between interaction strength and the energetic flux across a predator–prey link has been poorly studied and would appear to offer considerable potential for future food-web research (Closs, 1996; Benke and Wallace, 1997; Goldwasser and Roughgarden, 1998). Our examination of the dynamics of generalist predators suggests that interaction strength and energetic flux may be related to some degree, and that both may vary significantly spatially and temporally. Comparison between theoretical conjecture and empirical reality is, however, severely hampered by a lack of food-web studies that have attempted to measure interaction strength and energetic flux simultaneously, the exceptions being the pioneering studies by Paine (1980, 1992). Raffaelli and Hall (1992) and Closs (1996) suggested that weak interactions dominated following predator manipulations in communities from which food webs had been previously described (Closs and Lake, 1994; Raffaelli and Hall, 1996). Detailed measurement of energetic flux across links, however, were not made in either case. Several other food-web studies have used a variety of methods to estimate energetic flux but have provided no measure of interaction strength (Goldwasser and Roughgarden, 1993; Tavares-Cromar and Williams, 1996; Benke and Wallace, 1997).

Our analysis suggests that spatial and temporal changes in the dynamics of interaction strength and energetic flux across trophic links are most likely to occur in the period immediately following the introduction or removal of a species. Surprisingly few studies have examined spatial and temporal changes in interaction strength and energetic flux following predator introduction or removal from a community. Oscarson (1987) documented changes in the distribution of prey and diet of fish following the introduction of roach to a formerly fish-free pond. In the period immediately following their introduction, the diet of roach was dominated by an open-water species of corixid. This species soon became extinct, after which roach switched to feeding on a second species of corixid that was largely restricted to littoral weed beds. Changes in the diet of predators following their introduction to formerly predator-free island habitats also appears to follow a similar pattern. A typical example is that described by Savidge (1987), where the introduced brown tree-snake *Boiga irregularis* Merrem switched to a diet of rats and lizards after eliminating vulnerable species of native bird on the island of Guam. Numerous similar examples exist in the literature (Closs *et al.*, in preparation).

B. Top-down and Bottom-up Regulation and the Role of Refugia

Our examination of generalist predation in space and time suggests a key role for refugia in determining the relative importance of top-down and bottom-up regulation of food webs. We argue that the availability of refugia and resources within those refuges are likely to be key determinants of prey production, particularly in habitats where predators are ubiquitous. The significance of refugia as stabilising factors is somewhat uncertain with theoretical and empirical studies of specialist predator–prey systems suggesting that refugia have the potential to either stabilise or destabilise interactions in at least simple systems (McNair, 1986; Murdoch *et al.*, 1996). Empirical studies of more complex systems have demonstrated that refugia can play an important role in permitting co-existence of predators and prey (see reviews by Jeffries and Lawton, 1984; Diehl, 1995; Diehl and Eklov, 1995; Dudgeon, 1996). However, to our knowledge no studies have examined whether there is a quantitative relationship between refuge availability, prey production and subsequent predator production. The observation that resource limitation within spatial refugia can result in reduced rates of prey growth (Mittelbach, 1988; Diehl and Eklov, 1995) suggests that refuge size and quality has the potential to influence subsequent rates of predator production. The study of this relationship has significant applied implications given that the increasing advocacy and use of "no-take" zones as a fisheries management tool rests on the assumption that increased refuge space results in sustained prey production irrespective of fishing pressure outside the refuge (Joshi and Gadgil, 1991; Schmidt, 1997).

Related to the role of refugia in stabilising predator–prey population dynamics is the idea that the impact of predation on any particular prey generation will have little influence on the abundance of subsequent generations. We argue that in communities dominated by non-dynamic strong and weak interactions, there will be little or no relationship between the impact of predation and the abundance of subsequent prey generations. Predators will consume surplus prey production, whilst the numbers of prey individuals actually surviving to breed will be determined by refuge availability and quality. Weak coupling between predation at any particular site and patterns of prey recruitment in subsequent generations has clear parallels with aspects of supply-side population dynamics that have been described for marine intertidal systems, where the population of prey at a particular site is determined by the supply of recruits from elsewhere (see Underwood and Fairweather, 1989).

A similar situation may also occur in streams. Studies of the relationship between prey production and fish consumption suggest that fish may consume a substantial proportion of total prey production (Allen, 1951; Huryn, 1996), yet prey abundance remains high. The observation of high levels of genetic differentiation at very small spatial scales in mayflies (Schmidt *et al.*, 1995) and the high fecundity of many adult stream invertebrates (Wilzbach and Cummins, 1989) suggests that the replenishment of prey populations might be occurring through the successful breeding of a relatively small number of surviving individuals from each generation that disperse widely. If so, there is likely to be very weak coupling of predator–prey population dynamics from generation to generation.

Lotka–Volterra models of predator–prey population dynamics typically assume predator–prey population dynamics are strongly coupled across generations. Few studies of predator–prey dynamics have, however, examined the degree of coupling of population dynamics across generations. Again, this relationship has important implications for the management of pest or introduced generalist predators. Our model of generalist predation suggests that, once vulnerable prey species have been eliminated, the strength of interactions between the predator and remaining prey species will be weak and continued predation will not necessarily result in the local extinction of any additional species.

C. Species Richness and Food-web Stability

The description of increasing numbers of complex species-rich food webs dominated by weak interactions (e.g. Sprules and Bowerman, 1988; Warren, 1989; Winemillar, 1990; Hall and Raffaelli, 1991; Martinez, 1991, 1992; Polis, 1991; Paine, 1992; Raffaelli and Hall, 1992, 1996; Closs and Lake, 1994; Tavares-Cromar and Williams, 1996) would appear to suggest that dynamic constraints play little or no role in determining food-web structure under most circumstances (Lawton, 1992, 1995). The dynamics of individual interactions may, however, strongly influence overall web structure in simple food webs, thus permitting L–V dynamics to dominate. Compelling examples of systems clearly driven by reciprocal regulation of predator and prey populations have been reported from protist (Harrison, 1995), *Daphnia*-algae (McCauley and Murdoch, 1987, 1990), fish-cladocera-algae (Cryer *et al*., 1986), and small-mammal systems (Hansson, 1987; Reid *et al*., 1997). The observation that such dynamics are potentially highly unstable (Harrison, 1995; McCauley and Murdoch, 1987, 1990) suggests that long food chains in webs containing few species (see Bengtsson and Martinez, 1996) will be less likely to occur when compared with food-chain length in species-rich, highly reticulate webs.

Mesocosm studies in which communities of increasing complexity are assembled (see Allen-Morley and Coleman, 1989; Jenkins *et al*., 1992; Lawler and Morin, 1993; Morin and Lawler, 1996; Spencer and Warren 1996a,b) have suggested that relationships between food-web structure, stability and energy availability exist, but are more complex than either simple L–V or D-C models predict. Such experimental mesocosms will continue to offer considerable potential in teasing apart relationships between web complexity and stability such as those predicted by our model of generalist predation (see Kareiva, 1989; Holyoak and Lawler, 1996). In particular, exploration of the stability properties of food webs as increasing numbers of potential prey are progressively added, and the mechanisms that might stabilise individual interactions within food webs may well be highly rewarding. The model of generalist predation outlined previously has focused on the stabilising properties of spatial refugia; however, a wide range of more transient forms of refugia exist, including prey and patch switching by predators as prey availability declines (Murdoch and Oaten, 1975; Murdoch, 1979; Townsend *et al*., 1986), metapopulation dynamics (Holyoak and Lawler, 1996) and predator avoidance by prey (Sih, 1987).

D. Theoretical Models

Theoretical models of food webs are inevitably simple caricatures of the complexity of real systems (Pimm *et al.*, 1991). Current dynamic models are represented by the existing L–V model which assumes dynamic, uniformly strong interactions, or the D-C model which assumes that predators have zero impact on the population dynamics of their prey (Pimm, 1982; Lawton, 1989). As such, the existing models represent the two extremes of the continuum in interaction strength (Figure 1; Lawton, 1989). We concur with Pimm *et al.* (1991) in that simple models that make clear predictions are more likely to advance ecological theory than highly complex models with dynamics as complex as real food webs. Existing food-web models have proved hugely successful in generating straightforward hypotheses that can be compared with reality; however, we must always be prepared to consider "better" models (Pimm *et al.*, 1991). In particular, we suggest that there is a need to develop models of generalist predator–prey dynamics that incorporate mechanisms that permit variation in interaction strength in space and time, and explore stability properties of generalist predation models as additional prey species are added (Closs *et al.*, in preparation). Current models of food-web and generalist predator dynamics assume unvarying interaction strength. As discussed previously, abundant empirical evidence suggests that interaction strength between a single generalist predator and its prey does vary in space, time and with respect to prey species. Existing models fail to reflect the diversity in interaction strength that may lie at the core of food-web stability.

VIII. CONCLUSIONS

The search for emergent patterns that might stabilise complex food webs (e.g. Pimm, 1982) may have diverted attention from a key property of the Lotka–Volterra equations, i.e. that individual predator–prey interactions based purely on such dynamics are inherently unstable (Murdoch and Oaten, 1975). In relatively simple webs based on such dynamics, populations of component species are likely to fluctuate wildly and may go extinct (Murdoch and Oaten, 1975). It is difficult to envisage how complex webs based on equilibrium, reciprocal regulation of predator and prey populations could persist in a highly variable world whilst so delicately balanced on "a knife-edge of stability" (Murdoch and Oaten, 1975; Murdoch, 1991). This would appear to be particularly so for webs dominated by generalist predators. Numerous mechanisms that stabilise predator–prey interactions have been described including spatial refuges, invulnerable size-classes of prey and prey behavioural responses and prey switching by predators (DeAngelis and Waterhouse, 1987; Kerfoot and Sih, 1987; Hastings, 1988; Begon *et al.*,

1996b). Conceivably, stability could be generated at multiple levels of spatial and temporal scale within a habitat in which many different predators and prey species are likely to be interacting within a complex web, and in which numerous forms of enemy-free space are likely to be available (Jeffries and Lawton, 1984; Gonzalez and Tessier, 1997). An absence of dynamic constraints on overall food-web structure also suggests a greater role for energetics and the nature of local species pools in shaping food webs (see Jenkins *et al.*, 1992).

Constant dialogue between theory and real data is critical for the development of our understanding of food-web dynamics (Kareiva, 1989; Lawton, 1992; Harrison, 1995; Holyoak and Lawler, 1996). Existing L–V and D-C food-web models have formed a fruitful basis for theoretical conjecture and empirical testing over the past 20 years. We believe, however, that a growing body of empirical evidence suggests that neither model adequately represents the dynamics of complex multi-species food webs (see also Lawton; 1992; Fagan 1997). It would be unfortunate, however, if the apparent inadequacy of current models of food-web dynamics failed to motivate attempts to develop new models that aim to explain food-web stability and persistence (see also Lawton, 1992; Polis *et al.*, 1996). The opposing theoretical conjectures of Elton (1927) and MacArthur (1955) of "complexity begets stability" on the one hand, and May's (1972) observation that complexity results in instability, remain unresolved in any integrated sense and offer a fascinating challenge. As we have argued through this paper, we believe that an understanding of the dynamics of generalist predation can reconcile several key aspects of food-web dynamics, in particular the nature of interaction strength, relationships between interaction strength and energetic flux, and the persistence of complex, reticulate food webs through time. Closer examination of the dynamics of generalist predators suggests the potential for models that are rich in dynamical properties and that more closely reflect the properties of real webs.

ACKNOWLEDGMENTS

Comments by Colin Townsend, Ross Thompson, Sam Lake, Dave Raffaelli and Glassov Speights greatly improved this paper. Conversations with numerous colleagues and students over several years, particularly at Monash and La Trobe Universities, helped in the development of many of the ideas expressed here. Many thanks to you all.

REFERENCES

Abrams, P.A. (1993). Why predation rate should not be proportional to predator density. *Ecology* **74**, 726–733.

Allan, J.D. (1995). *Stream Ecology*. Chapman and Hall, London.

Allen, K.R. (1951). The Horokiwi stream: A study of a trout population. *New Zealand Mar. Dept Fish. Bull.* **10**, 1–238.

Allen-Morley, C.R. and Coleman, D.C. (1989). Resilience of soil biota in various food webs to freezing perturbations. *Ecology* **70**, 1127–1141.

Andersson, M. and Erlinge, S. (1977). Influence of predation on rodent populations. *Oikos* **29**, 591–597.

Bachman, R.A. (1984). Foraging behaviour of free-ranging wild and hatchery brown trout in a stream. *Trans. Am. Fish. Soc.* **113**, 1–32.

Bechara, J.A., Moreau, G. and Planas, D. (1992). Top-down effects of brook trout (*Salvelinus fontinalis*) in a boreal forest stream. *Can. J. Fish. Aquat. Sci.* **49**, 2093–2103.

Bechara, J.A., Moreau, G. and Hare, L. (1993). The impact of brook trout (*Salvelinus fontinalis*) on an experimental stream benthic community: The role of spatial and size refugia. *J. Anim. Ecol.* **62**, 451–464.

Begon, M., Mortimer, M. and Thompson, D.J. (1996a). *Population Ecology*. Blackwell Scientific Publications, Oxford

Begon, M., Harper, J.L. and Townsend, C.R. (1996b). *Ecology*. Blackwell Scientific Publications, Oxford.

Bengtsson, J. and Martinez, N. (1996). Causes and effects in food webs: Do generalities exist. In: *Food Webs* (Ed. by G.A. Polis and K.O. Winemillar), pp. 179–184. Chapman and Hall, New York.

Benke, A.C. and Wallace, J.B. (1997). Trophic basis of production among riverine caddisflies: Implications for food web analysis. *Ecology* **78**, 1132–1145.

Brusven, M.A. and Rose, S.T. (1981). Influence of substrate composition and suspended sediment on insect predation by the torrent sculpin (*Cottus rhotheus*). *Can. J. Fish. Aquat. Sci.* **38**, 1444–1448.

Closs, G.P. (1996). Effects of a predatory fish (*Galaxias olidus*) on the structure of intermittent stream pool communities in southeast Australia. *Aust. J. Ecol.* **21**, 217–223.

Closs, G.P. and Lake, P.S. (1994). Spatial and temporal variation in the structure of an intermittent-stream food-web. *Ecol. Monogr.* **64**, 1–21.

Closs, G.P. and Lake, P.S. (1996). Drought, differential mortality and the coexistence of a native and an introduced fish species in a south east Australian intermittent stream. *Environ. Biol. Fishes* **47**, 17–26.

Cohen, J.E., Briand, F. and Newman, C.M. (1990). *Community Food Webs*. Springer Verlag, Berlin.

Connell, J.H. and Slatyer, R.O. (1977). Mechanisms of sucession in natural communities and their role in community stability and organization. *Am. Nat.* **111**, 1119–1144.

Crawley, M.J. (1992). Population dynamics of natural enemies and their prey. In: *Natural Enemies* (Ed. by M.J. Crawley), pp. 40–89. Blackwell Scientific Publications, Oxford.

Crowder, L.B. and Cooper, W.E. (1982). Habitat structural complexity and the interaction between bluegills and their prey. *Ecology* **63**, 1802–1813.

Cryer, M., Peirson, G. and Townsend, C.R. (1986). Reciprocal interactions between roach (*Rutilus rutilus*) and zooplankton in a small lake: Prey dynamics and fish growth and recruitment. *Limnol. Oceanogr.* **31**, 1022–1038.

Dayton, P.K. (1971). Competition, disturbance and community organisation: The provision and subsequent utilization of space in a rocky intertidal community. *Ecol. Monogr.* **41**, 351–389.

DeAngelis, D.L. (1975). Stability and connectance in food-web models. *Ecology* **56**, 238–243.

DeAngelis, D.L. and Waterhouse, J.C. (1987). Equilibrium and non-equilibrium concepts in ecological models. *Ecol. Monogr.* **57**, 1–21.
DeRuiter, P.C., Neutel, A.M. and Moore, J.C. (1996). Energetics and stability in belowground food webs. In: *Food webs* (Ed. by G.A. Polis and K.O. Winemillar), pp. 201–210. Chapman and Hall, New York.
Diehl, S. (1995). Direct and indirect effects of omnivory in a littoral lake community. *Ecology* **76**, 1727–1740.
Diehl, S. and Eklov, P. (1995). Effects of piscivore-mediated habitat use on resources, diet, and growth of perch. *Ecology* **76**, 1712–1726.
Dudgeon, D. (1996). The influence of refugia on predation impacts in a Hong Kong stream. *Arch. Hydrobiol.* **138**, 145–159.
Elliot, J.M. (1994). *Quantitative Ecology and the Brown Trout.* Oxford University Press, Oxford.
Elton, C.S. (1927). *Animal Ecology*. Sidgewick and Jackson, London.
Erlinge, S. (1987). Predation and noncyclicity in a microtine population in southern Sweden. *Oikos* **50**, 347–352.
Errington, P.L. (1946). Predation and vertebrate populations. *Q. Rev. Biol.* **21**, 144–177, 221–245.
Fagan, W.F. (1997). Omnivory as a stabilizing feature of natural communities. *Am. Nat.* **150**, 554–567.
Fairweather, P.G. (1987). Experiments on the interaction between predation and the availability of different prey on rocky seashores. *J. Exp. Mar. Biol. Ecol.* **114**, 261–273.
Flecker, A.S. and Townsend, C.R. (1994). Community-wide consequences of trout introduction in New Zealand streams. *Ecol. Appl.* **4**, 798–807.
Gilinsky, E. (1984). The role of fish predation and spatial heterogeneity in determining benthic community structure. *Ecology* **65**, 455–468.
Goldwasser, L. and Roughgarden, J. (1993). Construction and analysis of a large Caribbean food web. *Ecology* **74**, 1216–1233.
Gonzalez, M.J. and Tessier, A.J. (1997). Habitat segregation and interactive effects of multiple predators on a prey assemblage. *Freshwater Biol.* **38**, 179–191.
Hall, S.J. and Raffaelli, D. (1991). Food web patterns: Lessons from a species-rich web. *J. Anim. Ecol.* **60**, 823–842.
Hall, S.J. and Raffaelli, D. (1993). Food webs: Theory and reality. *Adv. Ecol. Res.* **24**, 187–239.
Hansson, L. (1987). An interpretation of rodent dynamics as due to trophic interactions. *Oikos* **50**, 308–318.
Harrison, G.W. (1995). Comparing predator–prey models to Luckinbill's experiment with *Didinium* and *Paramecium. Ecology* **76**, 357–374.
Hartman, K.J. and Margraf, F.J. (1993). Evidence of predatory control of yellow perch (*Perca flavescens*) recruitment in Lake Erie, U.S.A. *J. Fish Biol.* **43**, 109–119.
Hastings, A. (1978). Spatial heterogeneity and the stability of predator–prey systems: Predator-mediated coexistence. *Theor. Popul. Biol.* **14**: 380–395.
Hastings, A. (1988). Food web theory and stability. *Ecology* **69**: 1665–1668.
Havens, K. (1992) Scale and structure in natural food webs. *Science* **257**, 1107–1109.
Holt, R.D. (1977). Predation, apparent competition, and the structure of prey communities. *Theor. Popul. Biol.* **12**, 197–229.
Holt, R.D. and Lawton, J.H. (1994). The ecological consequences of shared natural enemies. *Annu. Rev. Ecol. Syst.* **25**, 495–520.

Holyoak, M. and Lawler, S.P. (1996). Persistence of an extinction-prone predator–prey interaction through metapopulation dynamics. *Ecology* **77**, 1867–1879.
Hulsmann, S. and Mehner, T. (1997). Predation by underyearling perch (*Perca fluviatilis*) on a *Daphnia galeata* population in a short-term enclosure experiment. *Freshwater Biol.* **38**, 209–219.
Hunter, M.D. and Price, P.W. (1992). Playing chutes and ladders: Heterogeneity and the relative roles of bottom-up and top-down forces in natural communities. *Ecology* **73**, 724–732.
Hurlbert, S.H. (1997). Functional importance *vs* keystoneness: Reformulating some questions in theoretical biocenology. *Aust. J. Ecol.* **22**, 369–382.
Hurlbert, S.H. and Mulla, M.S. (1981). Impacts of mosquitofish (*Gambusia affinis*) predation on plankton communities. *Hydrobiologia* **83**, 125–151.
Huryn, A.D. (1996). An appraisal of the Allen paradox in a New Zealand trout stream. *Limnol. Oceanogr.* **41**, 243–252.
Jeffries, M.J. and Lawton, J.H. (1984). Enemy free space and the structure of ecological communities. *Biol. J. Linn. Soc.* **23**, 269–286.
Jenkins, B., Kitching, R.L. and Pimm, S.L. (1992). Productivity, disturbance and food-web structure at a local scale in experimental container habitats. *Oikos* **65**, 249–255.
Joshi, N.V. and Gadgil, M. (1991). On the role of refugia in promoting prudent use of biological resources. *Theor. Popul. Biol.* **40**, 211–229.
Kareiva, P. (1989). Renewing the dialogue between theory and experiments in population ecology. In: *Perspectives in Ecological Theory* (Ed. by J. Roughgarden, R.M. May and S.A. Levin), pp. 68–88. Princeton University Press, Princeton, NJ.
Kerfoot, W.C. and Sih, A. (1987). *Predation: Direct and Indirect Impacts on Aquatic Communities*. University Press of New England, London.
Lancaster, J. and Robertson, A.L. (1995). Microcrustacean prey and macroinvertebrate predators in a stream food web. *Freshwater Biol.* **34**, 123–134.
Lawler, S.P. (1993). Species richness, species composition and population dynamics of protists in experimental mesocosms. *J. Anim. Ecol.* **62**, 711–719.
Lawler, S.P. and Morin, P.J. (1993). Food web architecture and population dynamics in laboratory microcosms of protists. *Am. Nat.* **141**, 675–686.
Lawton, J.H. (1989). Food webs. In: *Ecological Concepts* (Ed. by J.M. Cherret), pp. 43–78. Blackwell Scientific Publications, Oxford.
Lawton, J.H. (1992). Feeble links in food webs. *Nature* **355**, 19–20.
Lawton, J.H. (1995). Webbing and WIWACS. *Oikos* **72**, 305–306.
Lawton, J.H. and Warren, P.H. (1988). Static and dynamic explanations for patterns in food webs. *Trends. Ecol. Evol.* **3**, 242–245.
Lima, S.L. and Dill, L.M. (1990). Behavioural decisions made under the risk of predation: a review and prospectus. *Can. J. Zool.* **68**, 619–640.
Lindstrom, E.R. (1994). Vole cycles, snow depth and fox predation. *Oikos* **70**, 156–160.
Macan, T.T. (1965). Predation as a factor in the ecology of water bugs. *J. Anim. Ecol.* **34**, 691–698.
Macan, T.T. (1966). The influence of predation on the fauna of a moorland fishpond. *Arch. Hydrobiol.* **61**, 432–452.
Macan, T.T. (1977). The influence of predation on the composition of fresh-water animal communities. *Biol. Rev.* **52**, 45–70.
MacArthur, R.H. (1955). Fluctuations of animal populations, and a measure of community stability. *Ecology* **36**, 533–536.

MacArthur, R.H. (1972). Strong, or weak, interactions? *Trans. Conn. Acad. Arts Sci.* **44**, 177–188.
McCauley, E. and Murdoch, W.W. (1987). Cyclic and stable populations: plankton as paradigm. *Am. Nat.* **129**, 97–121.
McCauley, E. and Murdoch, W.W. (1990). Predator–prey dynamics in environments rich and poor in nutrients. *Nature* **343**, 455–457.
McIntosh, A.R. and Peckarsky, B.L. (1996). Effects of trout and stoneflies on mayfly drift and positioning periodicity. *Freshwater Biol.* **35**, 141–148.
McIntosh, A.R. and Townsend, C.R. (1994). Interpopulation variation in mayfly antipredator tactics: Differential effects of contrasting predatory fish. *Ecology* **75**, 2078–2090.
McIntosh, A.R. and Townsend, C.R. (1995). Impacts of an introduced predatory fish on mayfly grazing in New Zealand streams. *Limnol. Oceanogr.* **40**, 1508–1512.
McNair, J.N. (1986). The effects of refuges on predator–prey interactions: A reconsideration. *Theor. Popul. Biol.* **29**, 38–63.
Martinez, N.D. (1991). Artifacts or attributes? Effects of resolution on the Little Rock Lake food web. *Ecol. Monogr.* **61**, 367–392.
Martinez, N.D. (1992). Constant connectance in community food webs. *Am. Nat.* **139**, 1208–1218.
Martinez, N.D. and Lawton, J.H. (1995). Scale and food-web structure—from local to global. *Oikos* **73**, 148–154.
May, R.M. (1972). Will a large complex system be stable? *Nature* **238**, 413–414.
May, R.M. (1973). *Stability and Complexity in Model Ecosystems.* Princeton University Press, Princeton, NJ.
May, R.M. (1977). Thresholds and breakpoints in ecosystems with a multiplicity of stable states. *Nature* **269**, 471–477.
Menge, B.A. (1992). Community regulation: Under what conditions are bottom-up factors important on rocky shores? *Ecology* **73**, 755–765.
Menge, B.A. and Lubchenco, J. (1981). Community organisation in temperate and tropical intertidal habitats: Prey refuges in relation to consumer pressure gradients. *Ecol. Monogr.* **51**, 429–450.
Menge, B.A., Lubchenco, J. and Ashkenas, L.R. (1985). Diversity, heterogeneity and consumer pressure in a tropical rocky intertidal community. *Oecologia* **65**, 394–405.
Menge, B.A., Berlow, E.L., Blanchette, C.A., Navarrate, S.A. and Yamada, S.B. (1994). The keystone species concept: Variation in interaction strength in a rocky intertidal habitat. *Ecol. Monogr.* **64**, 249–286.
Menge, B.A., Daley, B. and Wheeler, P.A. (1996). Control of interaction strength in marine benthic communities. In: *Food Webs* (Ed. by G.A. Polis and K.O. Winemillar), pp. 258–274. Chapman and Hall, New York.
Mittelbach, G.G. (1988). Competition among refuging sunfishes and effects of fish density on littoral zone invertebrates. *Ecology* **69**, 614–623.
Morin, P.J. and Lawler, S.P. (1996). Effects of food chain length and omnivory on population dynamics in experimental food webs. In: *Food Webs* (Ed. by G.A. Polis and K.O. Winemillar), pp. 218–230. Chapman and Hall, New York.
Murdoch, W.W. (1979). Predation and the dynamics of prey populations. *Fortschr. Zool.* **25**, 295–310.
Murdoch, W.W. (1991). The shift from an equilibrium to a non-equilibrium paradigm in ecology. *Bull. Ecol. Soc. Am.* **72**, 49–51.
Murdoch, W.W. (1994). Population regulation in theory and practice. *Ecology* **75**, 271–287.

Murdoch, W.W. and Bence, J. (1987). General predators and unstable prey populations. In: *Predation: Direct and Indirect Impacts on Aquatic Communities* (Ed. by W.C. Kerfoot and A. Sih), pp. 17–30. University Press of New England, London.
Murdoch, W.W. and Oaten, A. (1975). Predation and population stability. *Adv. Ecol. Res.* **9**, 2–132.
Murdoch, W.W., Chesson, J. and Chesson, P. (1985). Biological control in theory and practice. *Am. Nat.* **125**, 344–366.
Murdoch, W.W., Swarbrick, S.L., Luck, R.F., Walde, S. and Yu, D.S. (1996). Refuge dynamics and metapopulation dynamics: An experimental test. *Am. Nat.* **147**, 424–444.
Oksanen, L., Oksanen, T., Ekerholm, P., Moen, J., Lundberg, P., Schneider, M. and Aunapuu, M. (1996). Structure and dynamics of Arctic–Subarctic grazing webs in relation to primary productivity. In: *Food Webs* (Ed. by G.A. Polis and K.O. Winemillar), pp. 231–242. Chapman and Hall, New York.
Oscarson, H.G. (1987). Habitat segregation in a water boatman (Corixidae) assemblage—the role of predation. *Oikos* **49**, 133–140.
Paine, R.T. (1966). Food web complexity and species diversity. *Am. Nat.* **100**, 65–75.
Paine, R.T. (1980). Food webs: Linkage, interaction strength and community infrastructure. *J. Anim. Ecol.* **49**, 667–685.
Paine, R.T. (1988). Food webs: Road maps of interactions or grist for theoretical development? *Ecology* **69**, 1648–1654.
Paine, R.T. (1992). Food web analysis through field measurement of *per capita* interaction strength. *Nature* **355**, 73–75.
Pech, R.P., Sinclair, A.R.E., Newsome, A.E. and Catling, P.C. (1992). Limits to predator regulation of rabbits in Australia: Evidence from predator-removal experiments. *Oecologia* **89**, 102–112.
Persson, L. (1987). Effects of habitat and season on competitive interactions between roach (*Rutilus rutilus*) and perch (*Perca fluviatilis*). *Oecologica* **73**, 170–177.
Persson, L. and Greenberg, L.A. (1990a). Interspecific and intraspecific size class competition affecting resource use and growth of perch, *Perca fluviatilis*. *Oikos* **59**, 97–106.
Persson, L. and Greenberg, L.A. (1990b). Juvenile competitive bottlenecks: The perch (*Perca fluviatilis*)–roach (*Rutilus rutilus*) interaction. *Ecology* **71**, 44–56.
Persson, L., Bengtsson, J., Menge, B.A. and Power, M.E. (1996). Productivity and consumer regulation—concepts, patterns, and mechanisms. In: *Food Webs* (Ed. by G.A. Polis and K.O. Winemillar), pp. 396–434. Chapman and Hall, New York.
Pimm, S.L. (1982). *Food Webs*. Chapman and Hall, London.
Pimm, S.L., Lawton, J.H. and Cohen, J.E. (1991). Food-web patterns and their consequences. *Nature* **350**, 669–674.
Polis, G.A. (1991). Complex trophic interactions in deserts: An empirical critique of food web theory. *Am. Nat.* **138**, 123–155.
Polis, G.A. (1994). Food webs, trophic cascades and community structure. *Aust. J. Ecol.* **19**, 121–136.
Polis, G.A. and Strong, D. R. (1996). Food-web complexity and community dynamics. *Am. Nat.* **147**, 813–846.
Polis, G.A., Holt, R.D., Menge, B.A. and Winemillar, K.O. (1996). Time, space, and life history: Influences on food webs. In: *Food Webs* (Ed. by G.A. Polis and K.O. Winemillar), pp. 435–460. Chapman and Hall, New York.

Power, M.E. (1990). Effects of fish in river food webs. *Science* **250**, 811–814.
Power, M.E. (1992). Top-down and bottom-up forces in food webs: Do plants have primacy? *Ecology* **73**, 733–746.
Raffaelli, D.G. and Hall, S.J. (1992). Compartments and predation in an estuarine food web. *J. Anim. Ecol.* **61**, 551–560.
Raffaelli, D.G. and Hall, S.J. (1996). Assessing the relative importance of trophic links in food webs. In: *Food Webs* (Ed. by G.A. Polis and K.O. Winemillar), pp. 185–191. Chapman and Hall, New York.
Reid, D.G., Krebs, C.J. and Kenney, A.J. (1997). Patterns of predation on noncyclic lemmings. *Ecol. Monogr.* **67**, 89–108.
Rosemond, A.D., Mulholland, P.J. and Elwood, J.W. (1993). Top-down and bottom-up control of stream periphyton: Effects of nutrients and herbivores. *Ecology* **74**, 1264–1280.
Rosenzweig, M.L. and MacArthur, R.H. (1963). Graphical representation and stability conditions of predator–prey interactions. *Am. Nat.* **97**, 209–223.
Savidge, J.A. (1987). Extinction of an island forest avifauna by an introduced snake. *Ecology* **68**, 660–668.
Schmidt, K.F. (1997). "No-take" zones spark fisheries debate. *Science* **277**, 489–491.
Schmidt, S.K., Hughes, J.M. and Bunn, S.E. (1995). Gene flow among conspecific populations of *Baetis* sp. (Ephemeroptera): Adult flight and larval drift. *J. North Am. Benthol. Soc.* **14**, 147–157.
Schoener, T.W. (1989). Food webs: From the small to the large. *Ecology* **70**, 1559–1589.
Schoenly, K., Beaver, R.A. and Heumier, T.A. (1991). On the trophic relations of insects: A food web approach. *Am. Nat.* **137**, 597–638.
Sih, A. (1987). Predators and prey lifestyles: An evolutionary and ecological overview. In: *Predation* (Ed. by W.C. Kerfoot and A. Sih), pp. 203–224. University Press of New England, London.
Spencer, M. and Warren, P.H. (1996a). The effects of energy input, immigration and habitat size on food web structure: A microcosm experiment. *Oecologia* **108**, 764–770.
Spencer, M. and Warren, P.H. (1996b). The effects of habitat size and productivity on food web structure in small aquatic microcosms. *Oikos* **75**, 419–430.
Spiller, D.A. and Schoener, T.W. (1996). Food webs dynamics on some small subtropical islands: Effects of top and intermediate predators. In: *Food Webs* (Ed. by G.A. Polis and K.O. Winemillar), pp. 160–169. Chapman and Hall, New York.
Sprules, W.G. and Bowerman, J.E. (1988). Omnivory and food chain length in zooplankton food webs. *Ecology* **69**, 418–426.
Strong, D.R. (1992). Are trophic cascades all wet? Differentiation and donor-control in speciose ecosystems. *Ecology* **73**, 747–754.
Sugihara, G., Schoenly, K. and Trombla, A. (1989). Scale invariance in food web properties. *Science* **245**, 48–52.
Tavares-Cromar, A.F. and Williams, D.D. (1996). The importance of temporal resolution in food web analysis: Evidence from a detritus based stream. *Ecol. Monogr.* **66**, 91–113.
Thorp, J.H. (1986). Two distinct roles for predator in freshwater assemblages. *Oikos* **47**, 75–82.
Townsend, C.R., Winfield, I.J., Peirson, G. and Cryer, M. (1986). The response of young roach *Rutilus rutilus* to seasonal changes in abundance of microcrustacean prey: A field demonstration of switching. *Oikos* **46**, 372–378.

Underwood, A.J. and Fairweather, P.G. (1989). Supply-side ecology and benthic marine assemblages. *Trends Ecol. Evol.* **4**, 16–20.

Ware, D.M. (1972). Predation by rainbow trout (*Salmo gairdneri*): The influence of hunger, prey density and prey size. *J. Fish. Res. Bd. Can.* **29**: 1193–1201.

Warren, P.H. (1989). Spatial and temporal variation in the structure of a freshwater food web. *Oikos* **55**, 299–311.

Wilzbach, M.A. and Cummins, K.W. (1989). An assessment of short-term depletion of stream macroinvertebrate benthos by drift. *Hydrobiologia* **185**, 29–39.

Winemillar, K O. (1989). Must connectance decrease with species richness? *Am. Nat.* **134**, 960–968.

Winemillar, K.O. (1990). Spatial and temporal variation in tropical fish trophic networks. *Ecol. Monogr.* **60**, 331–367.

Zaret, T.M. (1980). *Predation and Freshwater Communities.* New Haven, Yale University Press.

Delays, Demography and Cycles: A Forensic Study

W.S.C. GURNEY, S.P. BLYTHE AND T.K. STOKES

I. SUMMARY

Strict logic implies that a hypothesis can only be proved wrong, but the main motivation of much science is to find out what is true. In this paper, we argue that the process of intuition building works by extending the classical Popperian paradigm of hypothesis and refutation to include a close relative of the legal concept of "reasonable doubt". We believe that the way in which doubts are eroded involves building up a consistent picture over a wide ranging body of data—a process akin to the forensic investigations common in the legal process. We illustrate our argument by re-analysing A.J. Nicholson's classical data set on the Australian sheep blowfly *Lucilia cuprina.* We conclude by discussing the implications of the investigative paradigm we describe for the design of experimental and observational programmes.

II. INTRODUCTION

One of many intriguing paradoxes in the practice of science is that, from a strictly logical perspective, it is only possible to *prove* a given hypothesis to be untrue. Since the range of plausible alternatives is invariably large enough to preclude exhaustive exploration, this implies that we can only be sure when

ADVANCES IN ECOLOGICAL RESEARCH VOL. 28
ISBN 0–12–013928–6

our view of the way some natural system functions is wrong. However, most people do science in order to find out how some piece of the world works, not to compile a list of incorrect explanations. Experienced practitioners generally develop very good intuition about how their piece of the world functions. However, they frequently give little consideration to the process by which such intuition is formed and, consequently, pay little attention to designing analytical and observational procedures which maximise the output of intuition.

In this paper we argue that the key to understanding how scientific understanding is achieved is to extend the idealised Popperian view of the scientific method as a process of hypothesis and falsification, to include the concept of "reasonable doubt" which is familiar in the legal process. This notion arises quite naturally if one regards the processes of experimentation and modelling/analysis as parts of an ongoing investigation, the aim of which is to build up a consistent picture of what is happening ecologically. A candidate model is defined, its parameters are estimated, and then its predictions are tested using independent data. Where it fails, a biologically plausible elaboration of the original model is proposed and tested. When this process produces a model consistent with a large body of data, we begin to feel that the model represents something more than an idealised view of reality which has not (yet) been proved wrong!

Two questions come to mind when considering this process. First, how large a body of data must it encompass before we can prudently feel any sort of confidence in the model. Secondly, what sort of data provide the best raw material. Neither question has a generic answer, so we base our discussion around the classic set of laboratory experiments on the Australian sheep blowfly *Lucilia cuprina* published by A.J. Nicholson (1954a,b, 1957). A subset of these data has recently been subjected to exhaustive re-analysis by Kendal *et al.* (1998), who investigated the utility of modern time-series techniques for model falsification, placing particular emphasis on the detection of false fits resulting from the combination of noisy data and structurally inappropriate models.

In contrast, we describe a historical process which used less sophisticated probes, but examined a wider ranging body of data. We begin with a model which successfully fitted data from an early short experiment, but was apparently falsified by later, more extensive experiments. We then show how confidence is enhanced by an ability to encompass the extended experimental results within conservative extensions of the original structure rather than by root and branch re-formulation. We conclude by examining the implications of this progressive confidence-building paradigm for the design of experimental and observational programmes.

III. THE 1954 EXPERIMENT

In the early 1950s, A.J. Nicholson performed a series of laboratory experiments, reported in Nicholson (1954a,b), designed to elucidate the role played by intraspecific competition in the population dynamics of the Australian sheep blowfly, *Lucilia cuprina.* Although these experiments examined competition between both larval and adult stages for a variety of resources, one has become a classic textbook illustration of population cycles (Varley *et al.*, 1973; Hastings, 1997).

In this experiment, the adults competed for a limited supply (500 mg day^{-1}) of ground liver, from which they derived the protein required for egg production. All other resources (including those needed by other life-history stages) were provided in excess. In Figure 1 we show the time-series of egg and adult numbers from the published replicate of this experiment. Its most immediately striking feature is the large "cyclic" fluctuations in both measured quantities. However, closer examination shows that these cycles have some surprising characteristics. First, the maximum egg population, and hence (since eggs take only about a day to hatch) the maximum rate of egg production, occurs when the adult population is near the minimum of its cycle. Secondly, each burst of egg production appears to have a characteris-

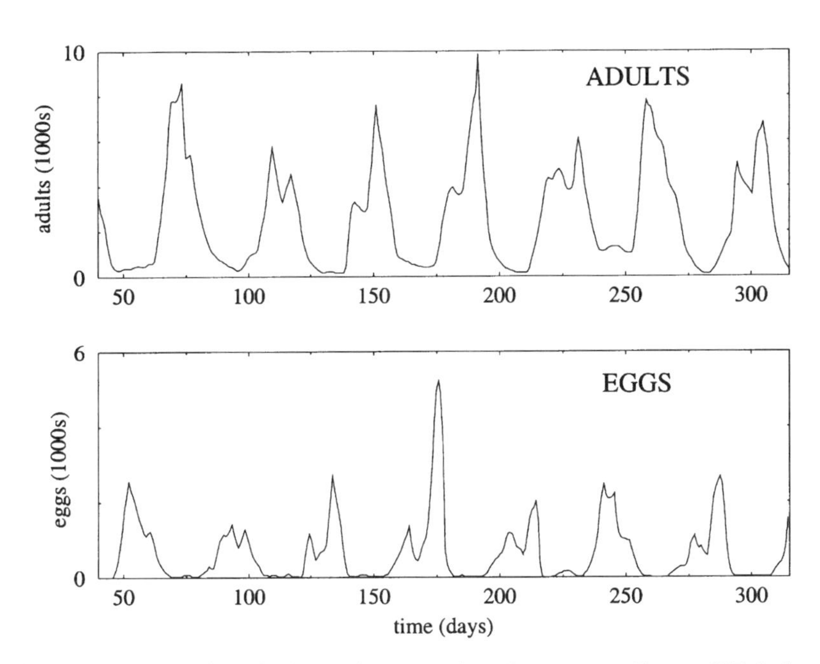

Fig. 1. The time series of adult and egg numbers from one replicate of Nicholson's 1954 experiment on *Lucilia cuprina,* in which adults competed for a limited supply (500 mg day^{-1}) of the protein (ground liver) needed to fuel egg production. Resources needed by all other life-history stages and all other resources needed by adults were provided *ad libitum.* Diagram replotted from data electronically recovered from the graphs published in Nicholson (1954b).

tic "double-peaked" form, which is reflected in the shape of the resulting peak in adult numbers.

Over the next three decades, a number of groups produced models which sought to elucidate the dynamics of this system, but all were either too parameter rich to be capable of falsification (Oster, 1977, 1981; Oster and Ipaktchi, 1978) or were empirical formulations weakly related to the underlying biology (May, 1976; Poole, 1979). In 1980, two groups independently produced models with a firm biological basis, yet simple enough to be rigorously tested using currently available information (Gurney *et al.*, 1980; Readshaw and Cuff, 1980). After a short, and at times robust, debate it was realised that these two models were structurally identical (see Gurney *et al.*, 1981; Readshaw, 1981). Over the next few years, this model, hereafter referred to as the Strathclyde-CSIRO model, was refined as further parameter information became available (Readshaw and Cuff, 1980; Readshaw and van Gerwen, 1983).

We set out a slightly simplified version of the treatment given by Gurney *et al.* (1983). All resources required by the larvae are supplied in excess, so the time (τ) taken to develop from egg to reproductively active adult, and the proportion (S) of individuals who successfully complete this transition, can be assumed to be constant. Since the experimental design rules out immigration and emigration, the adult population, $A(t)$, can change only through mortality and recruitment. If we write the per capita fecundity and mortality of the adults present in the culture at time t as $\beta(t)$ and $\delta(t)$, respectively, then the rate of egg production at that time is $\beta(t)A(t)$ and the rate of removal of adults by death is $\delta(t)A(t)$. However, adults recruited at time t result from eggs laid at time $t - \tau$, so the rate of change of the adult population at time t is

$$\frac{dA(t)}{dt} = S\beta(t-\tau)A(t-\tau) - \delta(t)A(t). \tag{1}$$

To complete our model, we have to select appropriate forms for the per capita vital rates β and δ, as well as evaluating the maturation delay, τ, and the egg to adult survival, S. The average age-distribution given by Nicholson (1954b), and redrawn here as the upper frame of Figure 2, shows us that

$$\tau = 15.5 \text{ days} \qquad S = 0.91. \tag{2}$$

Gurney *et al.* (1980) argued that mortality could be estimated from the rate of decline of the adult population during the parts of each cycle when recruitment is negligible. In the lower right-hand frame of Figure 2, we plot these declines on a logarithmic scale, and see that each shows a linear relation between ln A and t, thus indicating a constant per capita mortality during this

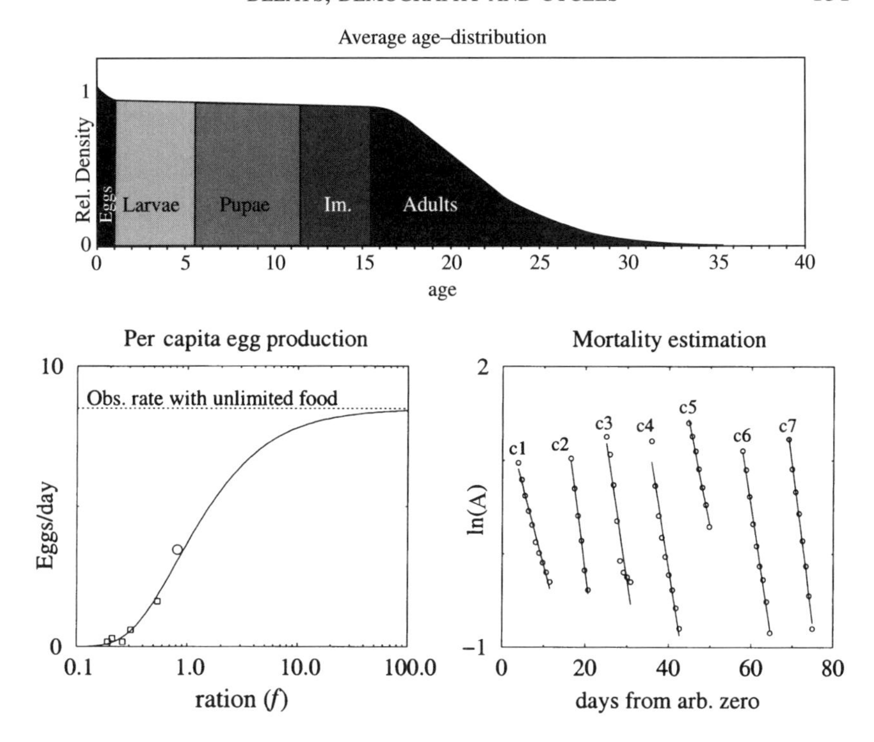

Fig. 2. Blowfly characteristics. Upper: average age distribution from Nicholson's cultures (redrawn from data in Nicholson (1954b). Lower left: egg production rate as a function of individual protein ration. Squares, data from Readshaw and Cuff (1980); circle, maximum per capita fecundity exhibited during the population runs; dotted line, upper limit deduced from data in Nicholson (1954b); solid line, fit given by equation (4). Lower right: adult population decline when recruitment negligible.

phase of the cycle. Although this mortality is quite variable from cycle to cycle, we see no evidence of systematic dependence on adult density, so we assume

$$\delta(t) = \delta_0 \tag{3}$$

where $\delta_0 = 0.27 \pm 0.02$ day^{-1}.

To determine the fecundity, Gurney *et al.* (1980) reanalysed the data set described in Readshaw and Cuff (1980), relating daily per-capita egg production of a caged blowfly population to individual protein ration (f). To these data, they added a further point, calculated from the maximum egg production observed in the population runs (assuming scramble competition), and an upper asymptote taken from the maximum possible fecundity observed by Nicholson (1954b). The composite data set was fitted (see the lower left-hand frame of Figure 2) by

$$\beta = \beta_{max} \exp\left[-\frac{f_0}{f}\right]. \tag{4}$$

where

$$\beta_{max} = 8.5 \text{ eggs day}^{-1}, \qquad f_0 = 0.833 \text{ mg day}^{-1}. \tag{5}$$

We next note that the culture is supplied with protein at a fixed rate, ϕ mg day^{-1}. Gurney *et al.* (1983) assumed scramble competition and complete consumption of supplied protein, so a population of A adults implies a daily ration of $f = \phi/A$ and a daily per capita egg production of $\beta(t) = \beta_{max}\exp[-A(t)f_0/\phi]$. Substituting this into equation (1) and defining

$$P = S\beta_{max}, \qquad A_0 = \phi/f_0, \tag{6}$$

leads to a compact statement of the Strathclyde-CSIRO blowfly model, namely

$$\frac{dA(t)}{dt} = R(t-\tau) - \delta A(t). \tag{7}$$

where

$$R(t) = PA(t)\exp\left[-\frac{A(t)}{A_0}\right]. \tag{8}$$

In Table 1 we summarise the parameter values for this model. We note that values for three out of the four (τ, P, A_0) were deduced from results independent of the population experiments against which we propose to test the

Table 1
Parameter values for the Strathclyde-CSIRO blowfly model

Parameter	Symbol	Value	Units
Maturation delay	τ	15.5	days
Maximum per capita reproduction	P	7.7	recruits/day
Characteristic population	A_0	600	adults
Adult mortality rate	δ_0	0.27	day^{-1}

model. The fourth (δ_0) was obtained by fitting to a distinct portion of each cycle, namely the period during which negligible recruitment leads to exponential population decline.

In Figure 3 we illustrate the quality of fit which these parameters imply to Nicholson's (1954) adult competition results. The starting condition of these cultures is not described in Nicholson's papers, so we have simply chosen the initial condition of the calculation so as to optimise the fit between data and predictions. However, it is important to realise that the initial condition does not affect the period, amplitude or shape of any cycle other than the first. Changing the initial condition thus only affects the phase (that is, the timing) of the majority of the predicted cycles.

The visual impression gained from Figure 3 is that the model fits the data extremely well. All the qualitative features of the data are present in the simulation. The adult and egg populations both exhibit large cyclic fluctuations. The majority of reproductive activity takes place during the periods of low adult population. This reproductive activity, and the adult peaks which result from it, both show a characteristic "double-peaked" shape. We also gain the

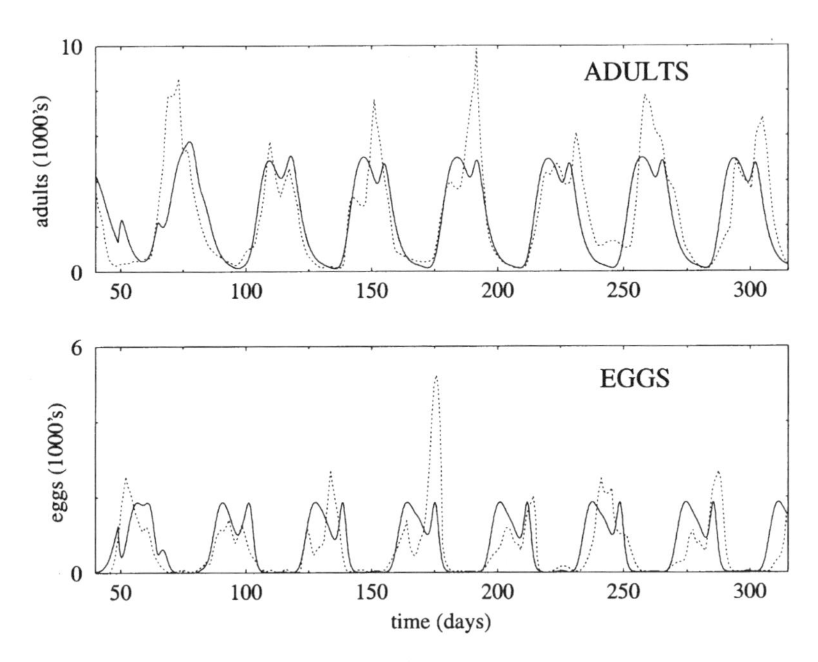

Fig. 3. The predictions of the Strathclyde-CSIRO blowfly model with the parameters given in Table 1 (solid line), compared to the data from Nicholson's (1954) adult competition experiment (dotted line). The initial condition, which does not affect the period, amplitude or shape of any of the cycles other than the first, has been chosen to optimise the coincidence between data and experiment.

impression that the quantitative predictions are close to the mark but not exactly right, and we confirm this in Table 2. This shows that the predicted cycle period is within the measurement uncertainty of that exhibited by the data, but the minimum and maximum populations are both somewhat underestimated.

The quality of this fit is extremely encouraging, especially since the parameters are mainly independently determined rather than being optimised. However, it is important to ask how robust this result is to the inevitable uncertainties in the parameter values. Gurney *et al.* (1980) showed that the parameters τ and A_0 are natural scales of time and population, respectively, and the behaviour of the system is really determined by the two parameter groups $P\tau$ and $\delta_0\tau$. In Figure 4 we show the response surface which they calculated. Superimposed on it we show the area covered if we assume that each parameter for which we have no uncertainty information is subject to a 10% measurement error. From this we see that the ratio of maximum to minimum population must lie in the region 25–50 and the ratio of period to delay must be in the range 2.35–2.45, suggesting that the minor error in maximum and minimum populations shown in Table 2 arises from a minor error in estimating the population scale A_0, that is from an error in the estimation of the characteristic ration f_0.

IV. THE 1957 EXPERIMENT

From a strictly Popperian perspective, the testing exercise described in the previous section has simply failed to falsify the Strathclyde-CSIRO model. However, the simplicity and biological plausibility of the model together with the qualitative and quantitative robustness of its predictions tempt us to believe that it may indeed represent some underlying biological "truth". It thus comes as a rude shock to discover that, in 1957, Nicholson published the results of an even longer and more intensive series of adult competition experiments (Nicholson, 1957), whose results show significant deviations from the predictions of the model.

Table 2
Strathclyde-CSIRO model predictions compared to the 1954 data

Quantity	Predicted	Observed	Units
Cycle period	37	38 ± 1.5	days
Maximum adult population	5400	7500 ± 500	adults
Minimum adult population	120	270 ± 120	adults
Double peak	Yes	Yes	—

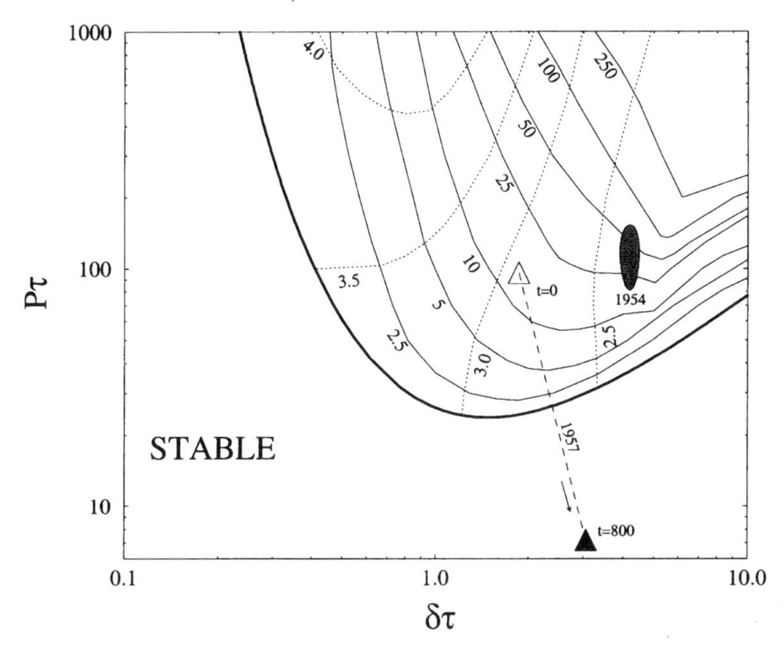

Fig. 4. Response surface for the Strathclyde-CSIRO blowfly model. Heavy solid line shows boundary between limit-cycles and stable equilibrium. Light solid lines are contours of constant A_{max}/A_{min}. Dotted lines are contours of constant period/delay. Shaded area shows parameter uncertainty for the 1954 experiment. Open and filled triangles show initial and final parameters, respectively, for the 1957 experiment. Redrawn from data recovered from Gurney *et al.* (1980).

In Figure 5 we show the adult numbers measured during an experiment in which the adults competed for a constant supply of 400 mg of ground liver per day. Over the first 400 days of the experiment the adult population shows large amplitude cyclic variations of similar form to those observed in the 1954 experiment, albeit with a rather less marked double-peaked form. However, in the second half of the experiment the population behaviour changes markedly. The visual impression of the new behaviour is that it is much less cyclic. This impression is reinforced by the periodograms in the lower two frames of the figure. The periodogram for the first 500 days of the run shows a well-marked peak at a period of 37 days (0.027 days^{-1}), but that for the last 200 shows a much broader area of cycle period running from as low as 30 to as much as 100 days (0.03 days^{-1} to 0.01 days^{-1}).

The impression that some quite marked change in behaviour takes place about a year into the experiment (approximately the duration of the 1954 experiment) is reinforced by the results of a second experiment which we show in Figure 6. Here the adult population competed for a supply of ground

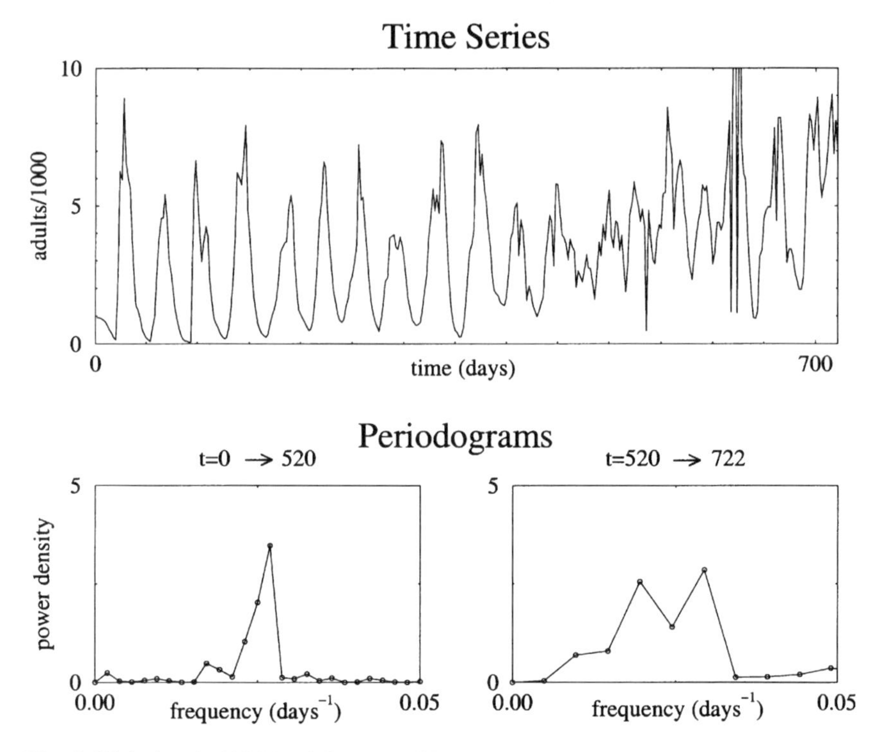

Fig. 5. Nicholson's (1957) adult competition experiment. Adults competed for a constant supply of 400 mg of ground liver per day. Upper frame shows adult population over time. Redrawn from data recovered from Nicholson (1957). Lower left and right frames show periodograms for $t = 0 \rightarrow 520$ and $t = 520 \rightarrow 722$, respectively. In both cases the trend was removed before the data were transformed.

liver which was varied cyclically between 500 mg day^{-1} and zero, with a period of 20 days. The periodograms shown in the lower frames of Figure 6 are the key to understanding what is going on. In the early part of the experiment, we see a very sharp peak at a period of 40 days (0.025 days^{-1}), which would be consistent with a limit cycle whose natural period was ≈ 37–38 days but which is synchronised to the first sub-harmonic of the variation in food supply rate (see Nisbet and Gurney, 1982). In the latter part of the experiment, the periodogram changes utterly, with all significant variation showing periods in the range 19–24 days.

Stokes *et al.* (1988) argued that such changes would be consistent with a dynamic system whose parameters changed during the experiment from values implying a deterministic limit cycle to values implying local stability. In the constant food experiment, such a change would imply cyclic fluctuations in the early part of the experiment, giving way in the latter part to random

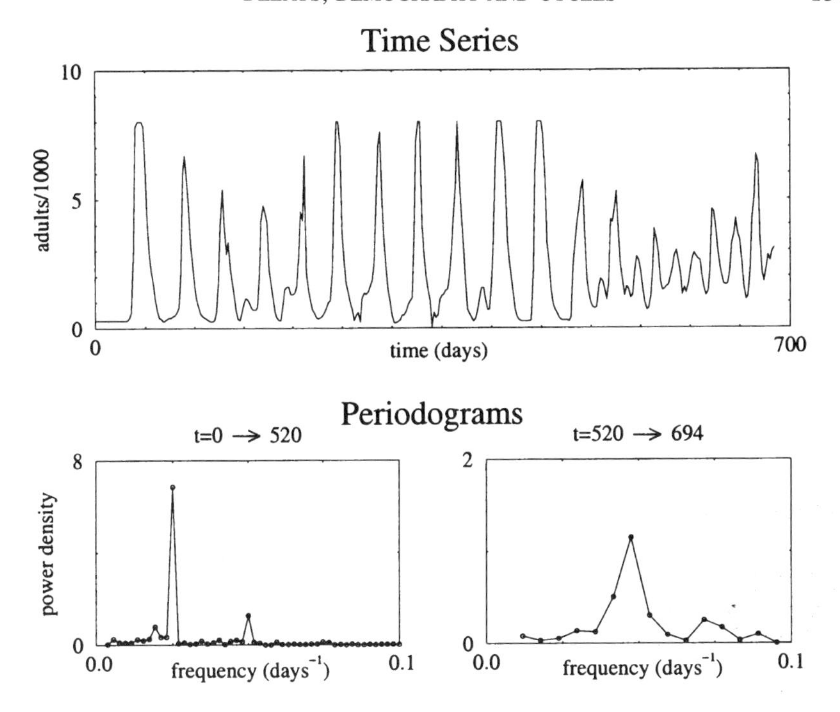

Fig. 6. Nicholson's (1957) adult competition experiment in which adults competed for a supply of ground liver which was varied cyclicly between 0 and 500 mg day^{-1} with a period of 20 days. Upper frame shows adult population over time. Redrawn from data recovered from Nicholson (1957). Lower left and right frames show periodograms for $t = 0 \rightarrow 520$ and $t = 520 \rightarrow 694$, respectively. In both cases the trend was removed before the data was transformed.

fluctuations as the stable steady-state is perturbed by demographic or environmental stochasticity. In the time-varying food experiment, the expected pattern would be cyclic population fluctuations in the early part of the experiment and, since their period is close to twice that of the food variation, it would be no surprise to find that they rapidly synchronise to this external "metronome". In the latter part of the experiment, the stable steady-state would be expected to "track" the variations in food supply rate, thus leading to cycles with a dominant period of around 20 days.

In addition to eggs and adults, Nicholson's (1957) data sets contain counts of the number of dead flies at each census. Stokes *et al.* (1988) were thus able to use the data from the constant food experiment to calculate time-resolved per capita mortality and fecundity rates. The resulting mortality rates showed high variability but only a very weak correlation with adult density. They were thus averaged over 100-day blocks to produce the time-varying parameter estimates shown in the lower frame of Figure 7. The fecundity estimates also showed very high variability, but this time well correlated with adult density. Hence Stokes *et al.* (1988) were able to analyse 100-day blocks of data to produce time-varying estimates of maximum reproduction rate, P, and characteristic ration, f_0 (see equation (4)), which we re-plot in the upper frames of Figure 7.

It is evident from Figure 7 that all the model parameters relating to adult characteristics (P, f_0 and δ_0) vary through the course of the experiment. To show that this parameter variation is responsible for the observed changes in system behaviour, Stokes *et al.* (1988) fitted smooth curves through the observed parameter values and plotted the resulting values on the response surface shown in Figure 4. The values start well inside the limit cycle region and end well inside the stable region, so we expect that, when we make a simulation run with this time-variation of parameters, the result will be a change in behaviour from limit cycles to stability around half-way through the run.

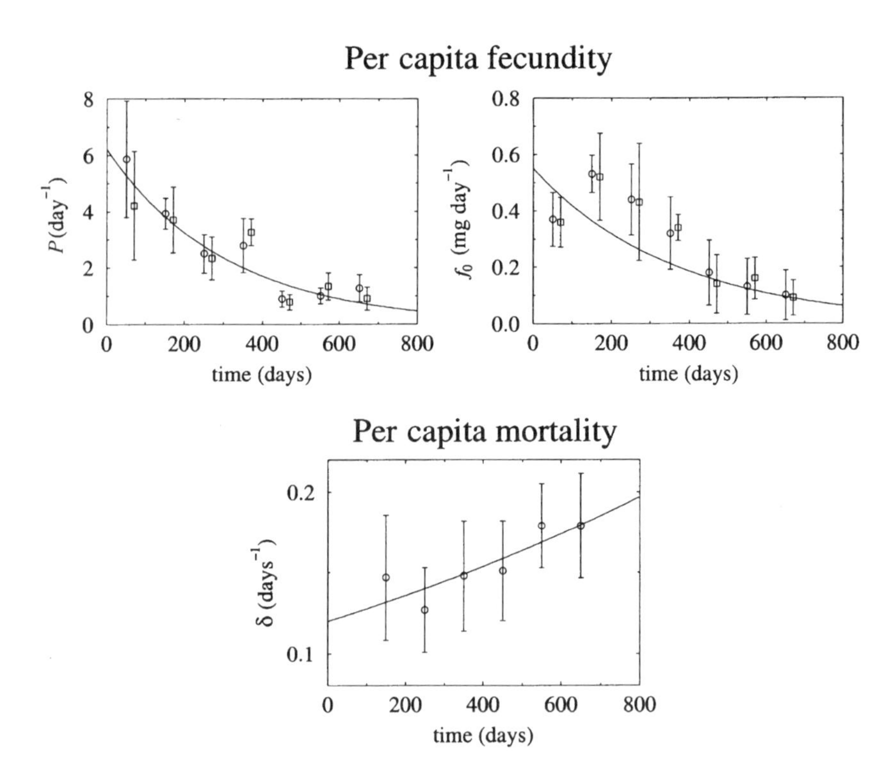

Fig. 7. Parameter variation inferred by Stokes *et al.* (1988) from Nicholson's (1957) constant food experiment. Solid lines show the fitted parameter variation used to create Figure 8. Redrawn using data recovered from graphs in Stokes *et al.* (1988).

Although the upper frame of Figure 8 confirms our expectations, we cannot view this as a true scientific test of the model. However, we can make such a test using the varying food data. In the lower frame of Figure 8, we show a run made with the same time variation of parameters, but now with the food supply rate, ϕ (see equation (6)), varying cyclically with a period of 20 days. The results are in remarkable concordance with the data. For the first half of the run the cycles are accurately phase-locked to twice the food-variation period and, in the second half, the 40-day period disappears and is replaced by a variation in which the adult population tracks the food input.

As a demonstration of the plausibility of parameter-drift as an explanation of the dynamics observed in the 1957 experiments, Stokes *et al.* (1988) made simulation runs with randomly varying parameters. In Figure 9 we present a related calculation, in which we use a model identical to that used to produce Figure 8, except that we regard the adult per capita mortality rate δ as a random variable. Our prescription for this variable is that it should be white noise, with a log-normal distribution (to prevent the occurrence of negative mortality!) and a power spectral density equal to that observed in the daily mortality rates from the constant food culture (0.02 days^{-1}). In the constant-

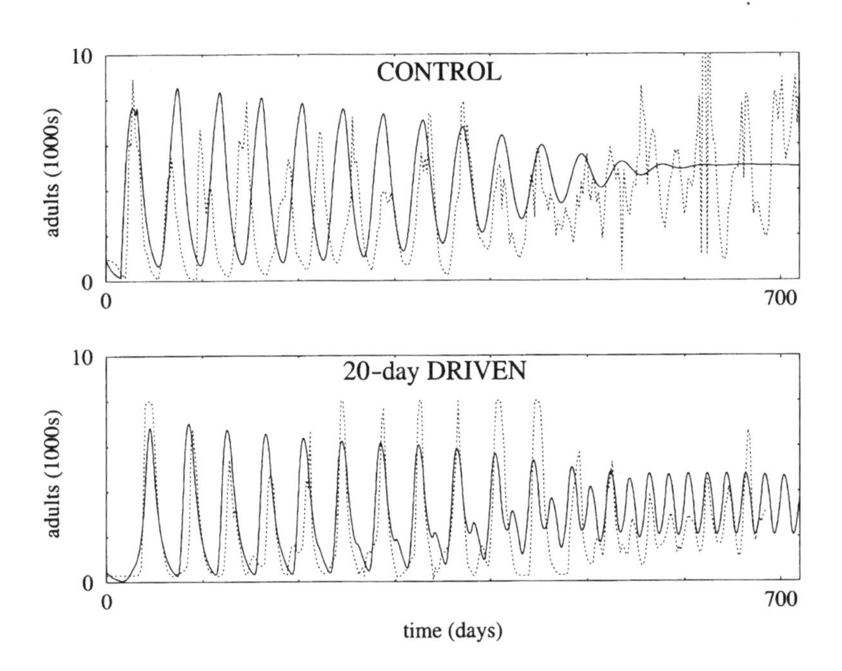

Fig. 8. Behaviour of the Strathclyde-CSIRO blowfly model with time-varying parameters deduced by Stokes *et al.* (1988) from the 1957 constant food experiment. Predictions are shown by the solid lines and data by the dotted lines. Upper frame, constant food experiment; lower frame, varying food experiment.

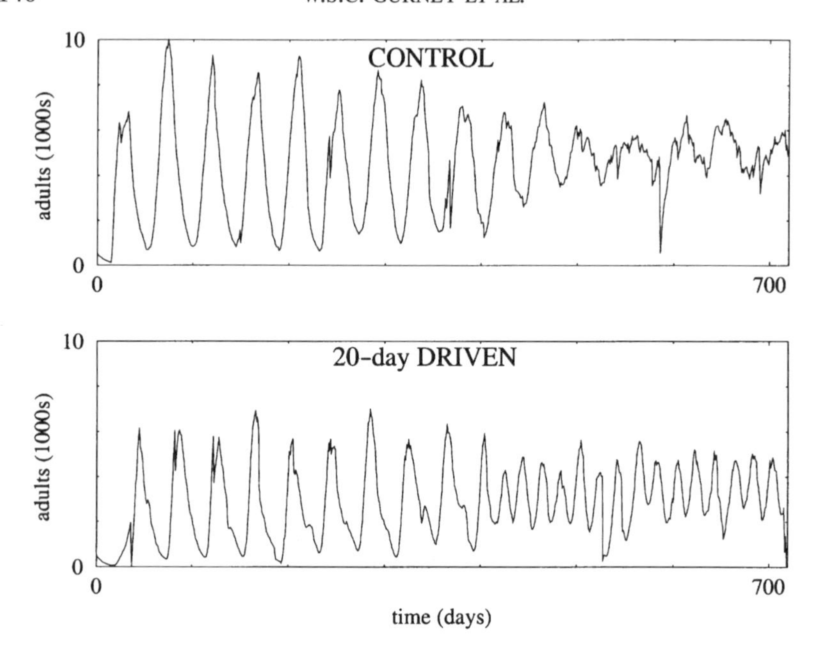

Fig. 9. Behaviour of a stochastic variant of the Strathclyde-CSIRO blowfly model with time-varying parameters and a randomly varying death rate in the form of log-normal distributed white noise with power spectral density 0.02 days^{-1}. Upper frame, constant-food experiment; lower frame, varying-food experiment.

food case, stochasticity produces little change in the cyclic portion of the run, but a plausibly random variability in its latter half. In the varying-food case, stochasticity in no way invalidates the concurrence shown by the deterministic simulation.

V. NATURAL SELECTION IN THE 1957 EXPERIMENT

Although the last section demonstrates that parameter drift provides a consistent explanation for the dynamic changes which occur during the 1957 experiments, it is natural to ask how such parameter changes can occur. Stokes *et al.* (1988) noted that the direction of the fecundity changes illustrated in Figure 7, that is reductions in both P and f_0 over time, are consistent with an alteration in the organism's performance which trades decreased fecundity at high food against increased reproductive ability at low food.

This change cannot be an altered phenotypic expression of a common genotype, since it would then appear as soon as the organism experienced the crowded conditions in the culture, that is pretty much as soon as the culture became established. By contrast, the changes in Nicholson's cultures happen over ten or more generations, a timescale which would be consistent with invasion by a different genotype with superior competitive abilities. If we regard the population parameters at the end of the experiment as typical of the invading genotype, then we see that it has sacrificed high food fecundity for the ability to reproduce successfully at low food, a behaviour occasionally observed in other species (Reynoldson, 1966). Such an adaptation would certainly be advantageous in the cyclic conditions of Nicholson's cultures, because an organism possessing it would start to reproduce sooner than its competitors as the population fell towards a trough and stop reproducing later as the population rose again.

Stokes *et al.* (1988) postulated that, at the start of the experiment, the population was dominated by a "pre-experimental type" fly with maximum reproductive rate P_w and characteristic ration f_w, but also contained a small number of "post-experimental type" animals with maximum reproductive rate P_e and characteristic ration f_e. By assuming that these two types cannot interbreed, it is possible to model the competition by a very simple extension of the Strathclyde-CSIRO model. Using R_w and R_e to denote the respective numbers of pre- and post-experimental type flies in the culture at time *t*, this extended model is

$$\frac{dA_w}{dt} = R_w(t-\tau) - \delta_w A_w, \qquad \frac{dA_e}{dt} = R_e(t-\tau) - \delta_e A_e, \tag{9}$$

where

$$R_w(t) = PA_w(t)\exp\left[-\frac{A_w(t)f_w + A_e(t)f_e}{\phi}\right] \tag{10}$$

and

$$R_e(t) = PA_e(t)\exp\left[-\frac{A_w(t)f_w + A_e(t)f_e}{\phi}\right]. \tag{11}$$

In Figure 10 we show the results of two simulations of this model, one in which the two groups of parameter values are determined from the average values over the first and last 100 days of the experiment, and the other in which they are inferred from the extremities of the smoothed fits through the measured data. In both cases we see the same outcome, namely that the pre-experimental type is eventually displaced by the post-experimental type, although the point of the experiment at which this happens is quite sensitively dependent on the relative competitive ability of the two types.

The "clonal" model proposed by Stokes *et al.* (1988) is assailable on the grounds that its central assumption, that the two types of fly cannot inter-

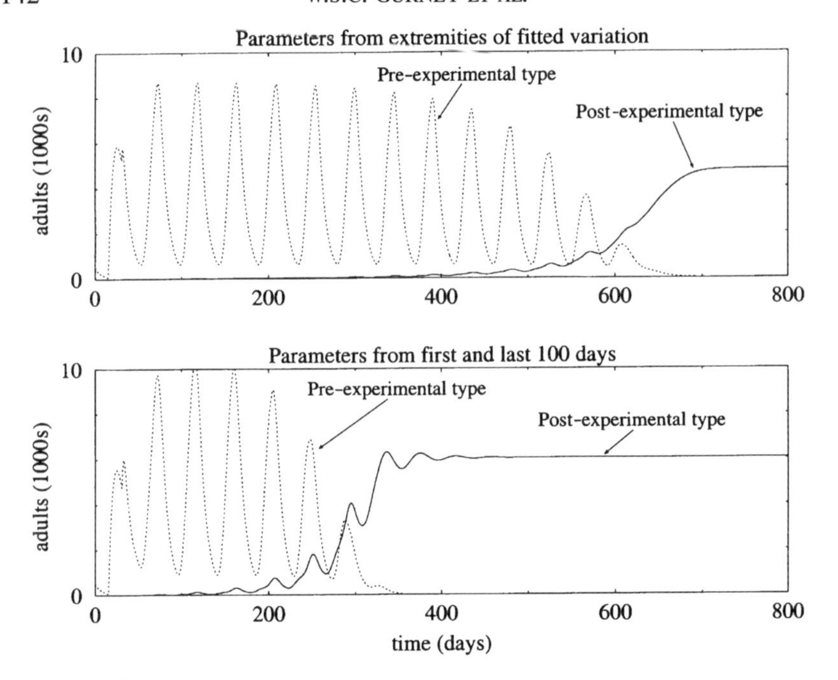

Fig. 10. Competition between pre-experimental type blowflies with parameters equal to population parameters at the start of the constant food run, and post-experimental flies with parameters equal to population parameters at the end of the run.

breed, is highly implausible. However, Nisbet *et al.* (1989) have modelled the invasion of an existing population by a new genotype, assuming a more conventional, one locus, two-allele, view of the genetic mechanisms. The conclusions from their study are broadly in line with the simplistic view resulting from the clonal model. With parameter differences of the size inferred for the blowfly case, the eventual outcome is always successful invasion, although its speed is greatly influenced by genetic details, such as whether the gene is dominant or recessive.

VI. DISCUSSION

Despite the logical impossibility of proving anything other than that a model is wrong, the foregoing sequence of discussion will lead many to a strong sense that the key elements in the Strathclyde-CSIRO model do indeed play a central role in the dynamics of Nicholson's laboratory blowfly populations. There are undoubtedly elements of the observed population variation the model cannot explain. For example, the observed double peaks are (almost certainly) systematically different in height, whereas the model predicts that they should normally be much the same height. It seems likely that this is a product of age-dependent adult mortality (see Readshaw and van Gerwen, 1983) and can thus only be explained by model incorporating adult age-structure. Similarly, there are clearly some features of the observations which are intrinsically stochastic in nature and cannot be explained by any deterministic model. None the less, the basic story, that the essential dynamic mechanism is the combination of a maturation delay of about 15 days with a per capita adult fecundity which drops strongly enough with competition for the population reproduction function[1] to have a "humped" form, by now seems extremely robust.

Although prudence as well as cold logic would rightly have inhibited such a conclusion after the early work on the 1954 experiment, we believe that it is now justified by the combined weight of favourable evidence provided by our ability to understand the more complex dynamics of the 1957 experiment within the same framework. Indeed, perhaps the evidential clincher is the way in which a relatively simple and biologically plausible framework proved capable of extension to provide a fully mechanistic explanation of the process of natural selection which produces these more complex dynamics.

This leads us to several more general points concerning the interaction of modelling and experimentation. First, we have demonstrated that a key element in the development of confidence in the correctness of the world view encapsulated by a particular model is its ability to encompass a large body of data rather than just the results of a single experiment. This seems to us to argue strongly that experiments which, in the absence of an underlying dynamic model, might seem to be merely minor extensions of observations already made can be key elements in the development of understanding.

Secondly, we conclude that an important aspect of confidence building is a clear understanding of the biological mechanisms being portrayed by the model. This argues strongly that simple and intuitive models will be more easily assimilated by this process than models with strongly counter-intuitive emergent behaviour. We support this contention by arguing that the initial apparent falsification of the Strathclyde-CSIRO model by the 1957 experiment was a vital step in the confidence-building process. This was partly because the model was simple enough to allow us to understand the dynamic changes which must underly the observations, thus paving the way for an extension of the model to explain what was going on. We note that such a process of refinement by falsification is very far removed from the construction of a model which seeks *ab initio* to explain every possible biological eventuality.

Lastly, we conclude that carefully constructed experimental observations are an essential part of the process. Although, viewed in retrospect, the 1957 constant-food experiment provided all the essential information about para-

[1]Cf. equation (8).

meter drift and natural selection, it was possible at the time to conceive of a number of more complex and exotic possibilities. The central evidence in ruling these exotica out of court was the food-varying experiment which acted as clear confirmation of the parameter drift hypothesis.

REFERENCES

Gurney, W.S.C., Blythe, S.P. and Nisbet, R.M. (1980). Nicholson's blowflies revisited. *Nature* **287**, 17–21.

Gurney, W.S.C., Blythe, S.P. and Nisbet, R.M. (1981). Reply to: The glass bead game. *Nature* **292**, 178.

Gurney, W.S.C., Nisbet, R.M. and Lawton, J.H. (1983). The systematic formulation of tractable single species population models incorporating age-structure. *Journal of Animal Ecology* **52**, 479–495.

Hastings, A. (1997). *Population Biology*. Springer Verlag, New York.

Kendall, B.E., Briggs, C.J., Murdoch, W.W., Turchin, P., Ellner, S.P., McCauley, E., Nisbet, R.M. and Wood, S.N. (1998). Understanding complex population dynamics: A synthetic approach. *Ecology* (in press).

May, R.M. (1976). Models for single species populations. In: *Theoretical Ecology: Principles and Applications* (Ed. by R.M. May). Blackwell, Oxford.

Nicholson, A.J. (1954a). Compensatory reactions of populations to stresses and their evolutionary significance. *Australian Journal of Zoology* **2**, 1–8.

Nicholson, A.J. (1954b). An outline of the dynamics of animal populations. *Australian Journal of Zoology* **2**, 9–65.

Nicholson, A.J. (1957). The self-adjustment of populations to change. *Cold Spring Harbour Symposium on Quantitative Biology* **22**, 153–173.

Nisbet, R.M. and Gurney, W.S.C. (1982). *Modelling Fluctuating Populations,* pp. 40–52. John Wiley & Sons, Chichester.

Nisbet, R.M., Gurney, W.S.C. and Metz, J.A.J. (1989). Stage structure models applied in evolutionary ecology. In: *Applied Mathematical Ecology* (Ed. by Levin, Hallam and Cross). Springer-Verlag, Dordrecht.

Oster, G. (1977). Internal variables in population dynamics. *Lectures in Mathematics in the Life Sciences* **8**, 37–68.

Oster, G. (1981). Predicting populations. *American Zoologist* **21**, 831–844.

Oster, G. and Ipaktchi, A. (1978). Population cycles. In: *Periodicities in Chemistry and Biology* (Ed. by H. Eyring). Academic Press, New York.

Poole, R.W. (1979). The statistical prediction of the fluctuations in abundance in Nicholson's sheep blowfly experiments. In: *Spatial and Temporal Analysis in Ecology* (Ed. by R.M. Cormak and K.J. Ord). International Co-operative publishing House Fairland, Maryland, USA.

Readshaw, J.L. (1981). The glass bead game. *Nature* **292**, 178.

Readshaw, J.L. and Cuff, W.R. (1980). A model of Nicholson's blowfly cycles and its relevance to predation theory. *Journal of Animal Ecology* **49**, 1005–1010.

Readshaw, J.L. and van Gerwen, A.C.M. (1983). Age-specific survival, fecundity and fertility of the adult blowfly *Lucilia cuprina* in relation to crowding, protein food and population cycles. *Journal of Animal Ecology* **52**, 879–887.

Reynoldson, T.B. (1966). The distribution and abundance of lake-dwelling Tri-clads: Towards a hypothesis. *Advances in Ecological Research* **4**, 1–71.

Stokes, T.K., Gurney, W.S.C., Nisbet, R.M. and Blythe, S.P. (1988). Parameter evolution in a laboratory insect population. *Theoretical Population Biology* **34**, 3, 248–265.

Varley, G.C., Gradwell, G.R. and Hassell, M.P. (1973). *Insect Population Ecology*. Blackwell, Oxford.

Spatial Root Segregation: Are Plants Territorial?

H.J. SCHENK, R.M. CALLAWAY AND B.E. MAHALL

I. SUMMARY

Spatially segregated root systems have been documented among conspecifics and among species at the scale of whole root systems and individual fine roots. Root segregation is often caused by architectural constraints, proliferation in particular microsites and plastic responses to competition for resources, but there is also evidence to suggest that allelopathy and non-toxic signals contribute to active root segregation. Root segregation appears to provide competitive advantages for water and nutrients for some species, as well as advantages of space itself. Plant growth and photosynthesis decreases when space is physically restricted, even when other resources are abundant. Moreover, plants appear to be able to compete for space independently of nutrient, water or light resources. Species that utilise resources efficiently and conservatively may particularly benefit from active root segregation because

ADVANCES IN ECOLOGICAL RESEARCH VOL. 28
ISBN 0-12-013928-6

more profligate neighbouring species would not be able to take resources that were being utilised slowly.

Stressful conditions, produced by adverse physical conditions and herbivory, have been shown to enhance the production of secondary metabolites and increase root exudation, mechanisms that can affect spatial root segregation. Resource availability may also determine the relative importance of root segregation in plant communities. A large portion of the evidence for root segregation comes from arid and semi-arid environments, where resources are often low. In resource-rich communities the defence of space may be less important.

For animals, the defence and exclusive use of space is considered to be evidence for territoriality and suggests that organisms that exhibit such behaviour are avoiding the costs and uncertainties of "scramble" competition. Active root segregation and the defence of space by plants indicates that plants also may be territorial and opens the possibility of a level of taxonomic generality in population biology that is not currently recognised.

II. INTRODUCTION

Plant roots often occupy soil volumes that are relatively free of other roots. Spatial segregation of root systems may lead to large competitive advantages because dominating belowground space promotes competitive ability (Brisson and Reynolds, 1997; Casper and Jackson, 1997). Segregation occurs among roots of the same species and among roots of different species, and has been referred to as root system antagonism (Bargioni, 1968), root exclusion (Nye and Tinker, 1977) and undermixing (Litav and Harper, 1967). Root segregation occurs among whole-root systems at the scale of metres and among individual roots at the scale of millimetres and, at both scales, segregation can have profound effects on competition for resources (Baldwin *et al.*, 1972; Escamilla *et al.*, 1991; Brisson and Reynolds, 1997). Roots spaced several centimetres apart may compete for mobile resources, but roots only millimetres apart may not compete for relatively immobile ones such as phosphate (Caldwell *et al.*, 1987; Beck *et al.*, 1989; Jungk, 1991).

Root segregation may be produced by architectural constraints and proliferation in preferred microsites and by active processes. In some cases roots appear to avoid resource depletion zones near other roots, indicating that belowground spatial segregation may be produced by competition. Other evidence indicates that roots may avoid soil volumes that are affected by the exudations of other roots, suggesting that root segregation may not always be directly related to resource uptake. Recent advances in methodology for studying roots (Callaway, 1990; Fitter and Stickland, 1991, 1992; Fitter *et al.*, 1991) and root interactions (Caldwell *et al.*, 1991, 1996; Mahall and Callaway, 1991, 1992, 1996; Brisson and Reynolds, 1994; Krannitz and

Caldwell, 1995) have produced new insights into belowground processes which indicate that spatial relationships among roots may be much more complex than previously recognised. Here, we review the literature on spatial root segregation and discuss the hypothesis that root segregation produced by the active defence of belowground space constitutes territoriality among plants.

III. MEASURING ROOT SEGREGATION

Methodological and analytical problems plague all studies of roots and have been discussed in a number of reviews (Böhm, 1979; Vogt *et al.*, 1989; Harper *et al.*, 1991; McMichael and Persson, 1991; Casper and Jackson, 1997). Excavation is a straightforward way to describe the natural distributions of roots (Coker, 1958; Bini and Chisci, 1961; Bargioni, 1962; Nobel and Franco, 1986; Brisson and Reynolds, 1994; Mou *et al.*, 1995). Excavation methods are effective for sampling large-scale patterns, but they often miss fine roots which are easily destroyed during digging. Alternatives to whole-root system excavation include plane-intersect methods, in which roots that intersect a cut surface in the soil are mapped (e.g. Weaver, 1919; Muller, 1946; Baldwin and Tinker, 1972; Escamilla *et al.*, 1991). In some plane-intercept approaches only roots that intersect a single surface are counted, whereas in others all roots that are exposed by washing away soil from a specified depth beneath the plant are counted (pinboard or bisect-wash methods). These methods allow precise measurements of all but the finest roots, but they are time consuming and generally are not used to describe whole-root systems of large plants.

Resolution of scale can be increased by deep-freezing soil cores in liquid nitrogen and cutting the cores into thin slices (Fusseder, 1983). However, the roots of different individuals and species are often difficult to distinguish. Litav and Harper (1967) labelled one of two neighbouring plants with ^{14}C and other authors have used ^{33}P, ^{32}P and ^{35}S labels in similar studies (Baldwin *et al.*, 1971; Baldwin and Tinker, 1972; Fusseder, 1983). Caldwell *et al.* (1991, 1996) distinguished shrub roots from grass roots at a very fine scale by their different fluorescence emissions under UV-A light.

Others have used root chambers or rhizotrons to detect root segregation. This technique permits the quantification of root distributions through time as they grow along a transparent surface. This non-destructive method allows estimation of root system dynamics at a very fine scale, although it creates unnatural growing conditions for roots growing along the viewing surface (Mackie-Dawson and Atkinson, 1991). The first study of root segregation to employ root chambers was Bergamini (1965) (Figure 1), who, however, did not analyse his results quantitatively. A variant of this method was developed by Mahall and Callaway (1991, 1992, 1996) and also used by Krannitz and

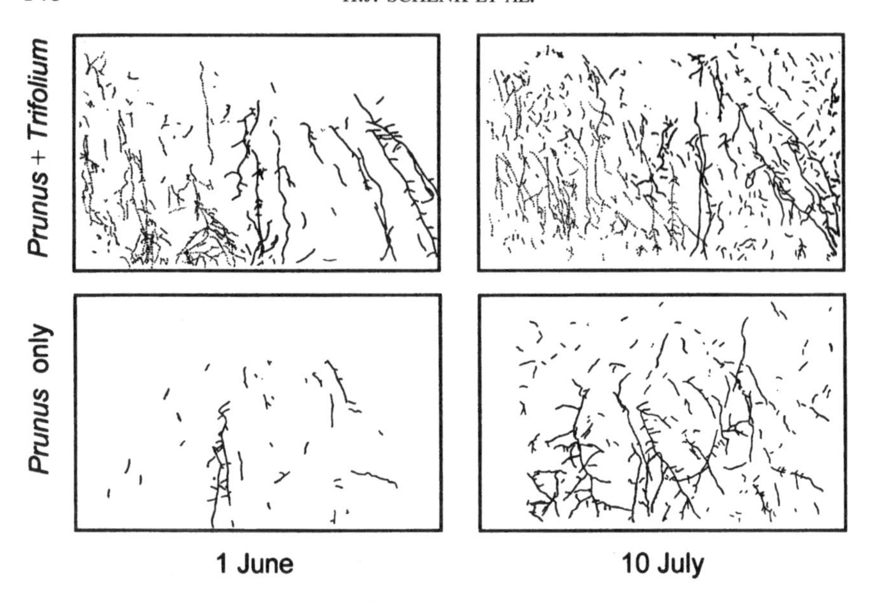

Fig. 1. Segregated root systems of *Prunus persica* (peach, solid lines) and *Trifolium pratense* (dotted lines) growing together in a root chamber. The pictures on the left side shows the distribution of roots growing along a transparent surface on 1 June 1965, the one on the left the distribution on 10 July 1965. Reproduced from Bergamini (1965).

Caldwell (1995). They grew the roots of a test plant into a root chamber containing a target plant rotated by 90° to increase root contacts (Figure 2). Nuclear magnetic resonance (NMR) and computerised image analysis has been used to map and analyse intact root systems (Sackville-Hamilton *et al.*, 1991; Southon and Jones, 1992).

Aboveground spatial patterns of plants have been used to estimate the extent of root system overlap indirectly, although this can lead to extensive errors (Brisson and Reynolds, 1997; Casper and Jackson, 1997). For example, chaparral shrubs occupy ten times more area below ground than above ground (Kummerow *et al.*, 1977), and *Schizachyrium scoparium* may root in an area that contains hundreds of other plants (Tilman, 1989). Aboveground positive associations have been found to occur among species with diverging root architectures (Cody, 1986a,b; Briones *et al.*, 1996), and regular distribution patterns within plant populations may indicate root system segregation, as has been hypothesised for *Larrea tridentata* (Fonteyn and Mahall, 1981; Mahall and Callaway, 1991). A reanalysis of data obtained by Gulmon and Mooney (1977) for two perennial species in Death Valley, California, *Atriplex hymenelytra* and *Tidestromia oblongifolia* by Upton and Fingleton (1985, pp.

Fig. 2. Picture of two connected root chambers containing *Larrea tridentata* plants. The roots from the test plant in the upper part of the picture have grown into the lower root chamber containing the target plant, rotated by 90° prior to connecting the two chambers to increase the probability of root contact. The markings indicate growth of the test-plant roots during 2-day intervals. Responses to root contact were measured by comparing elongation rates of test roots that contacted target roots with test roots that did not contact target roots.

257–258) showed that *Tidestromia* shrubs occupied areas that were free of conspecifics and which were referred to as territories. The belowground volumes that plants occupy have been estimated using relationships between plant size and interplant spacing, represented by Voronoi polygons drawn around individual plants (Mead, 1966; Mithen *et al.*, 1984; Welden *et al.*, 1990). The size of Voronoi polygons of the desert grass *Pleuraphis (Hilaria) rigida* was highly correlated with the volume occupied by root systems (Nobel, 1981). Bare spaces under plants have often been attributed to root mediated allelopathy and have been thought to indicate the exclusion of the roots of other plants (Went, 1955; Story, 1967; Jameson, 1970; Berlandier, 1980; Lange and Reynolds, 1981). These same kinds of patterns, however, have also been shown to be caused by volatile leaf allelochemicals (Muller *et al.*, 1964) and herbivory (Bartholomew, 1970).

IV. SEGREGATION OF WHOLE ROOT SYSTEMS

Aboveground spatial patterns of plant canopies have received considerable attention in the past (Harper, 1985; Jones and Harper, 1987a,b; Sprugel *et al.*, 1991; Hutchings and de Kroon, 1994) and, because roots are difficult to study, plant community theory has been based almost entirely on aboveground patterns (Fitter, 1987). However, belowground interactions are often the primary determinants of plant community structure. Many nutrient-uptake models are based on the assumption that roots of different plant species are distributed uniformly or at random with respect to each other (Nye and Tinker, 1977; Gregory, 1996), and spatially explicit vegetation and individual-based simulation models assume circular, symmetrical root distributions or "areas of influence" around individual plants (Firbank and Watkinson, 1985; Silander and Pacala, 1990; Mou *et al.*, 1993; Judson, 1994; Barth, 1995). The large-scale distribution of root systems, however, can be quite asymmetrical. Root distributions may correlate with aboveground patterns in some communities (Weaver, 1920; Kummerow *et al.*, 1977; Soriano *et al.*, 1987; Milchunas and Lauenroth, 1989; Hook *et al.*, 1994), but in others the correlation is weak (Weaver, 1920; Milchunas and Lauenroth, 1989; Lamont and Bergl, 1991). Roots of different species and individuals are often stratified vertically into well-defined layers and with highly variable horizontal patterns (Muller, 1946; Parrish and Bazzaz, 1976). Spatial stratification often appears to be due to genetically controlled differences in root architecture (Cannon, 1913; Parrish and Bazzaz, 1976; Cody, 1986b; Callaway, 1990), but root systems can be plastic, and spatial responses to heterogeneous conditions in the soil such as mechanical barriers, temperature, nutrients, water and oxygen contents vary widely within species (St John *et al.*, 1983; Caldwell and Richards, 1986; Coutts, 1989; Callaway, 1990; Fitter, 1994).

Other large-scale spatial segregation of roots appears to be produced by root exudates. In a few species, considered to be highly allelopathic (see section VI.B), root segregation has been found to occur at the scale of the whole plant, with very little overlap between neighbouring root volumes (Nye and Tinker, 1977). Most reported cases of large-scale root system segregation involve intraspecific spacing in plantations of crops and fruit trees (Table 1). Root systems of mature apple (*Malus domestica*) and persimmon (*Diospyros kaki*) trees show less horizontal overlap than would be expected from root distributions of isolated trees. Peach trees (*Prunus persica*) almost completely exclude roots of neighbouring conspecific trees from their rooting volume (Figure 3) (Bini and Chisci, 1961; Bargioni, 1962; Rogers and Head, 1969; Israel *et al.*, 1973; Nye and Tinker, 1977). Coker (1958) found that the degree

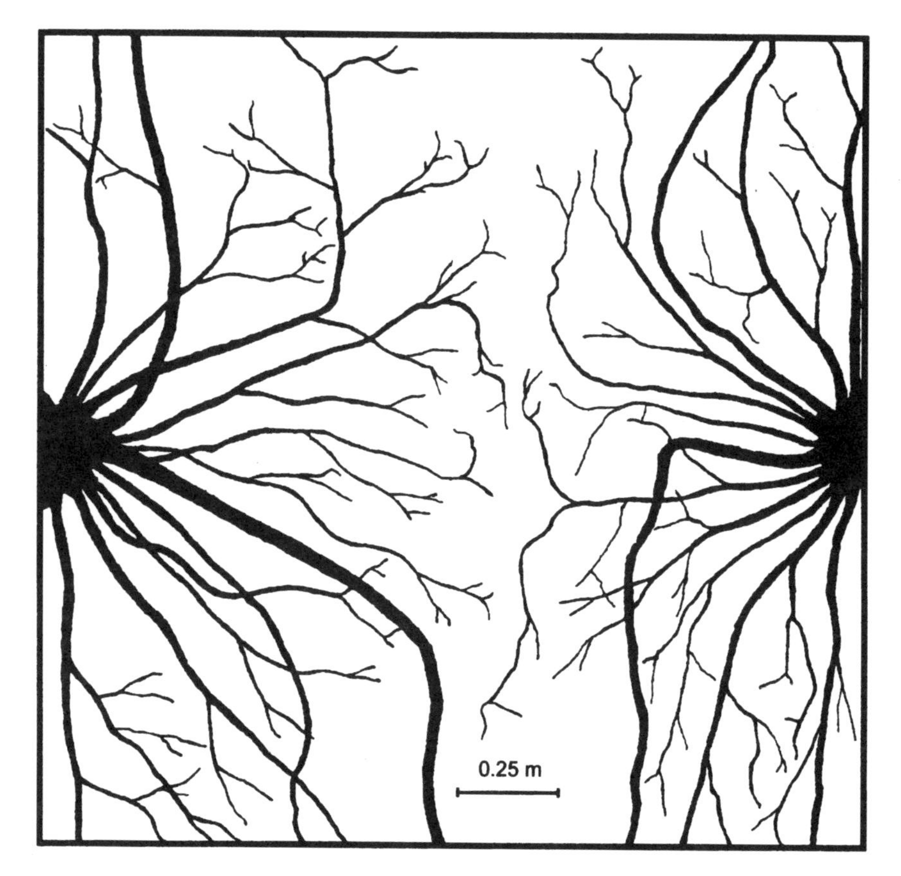

Fig. 3. Spatial distribution of the coarse-root system of two neighbouring peach (*Prunus persica*) trees. Modified from Bini and Chisci (1961).

Table 1

Evidence for intraspecific spatial segregation of root systems. Also listed are putative mechanisms, including toxic allelochemicals or signals, that have been suggested by various authors

Species	Method	Authors	Putative mechanism (e.g. allelochemical)
Allium cepa	Plane intersect, isotope labelling	Baldwin and Tinker (1972)	Unknown
Ambrosia dumosa	Root chamber	Mahall and Callaway (1991, 1992, 1996)	Unidentified signal, mediated by root contact (Mahall and Callaway, 1996)
Diospyros kaki	Excavation (whole root system)	Bargioni (1962)	Unknown
Glycine max	Excavation (pinboard method)	Raper and Barber (1970)	Unknown
Grevillea robusta	Excavation of seedlings	Webb *et al.* (1967)	Unidentified water-soluble toxin (Webb *et al.*, 1967)
Larrea tridentata	Excavation (whole root system); root chamber	Mahall and Callaway (1991, 1992); Brisson and Reynolds (1994)	Unidentified diffusable organic compound (Mahall and Callaway, 1992)
Liquidambar styraciflua	Excavation (whole root system)	Mou *et al.* (1995)	Unknown
Malus domestica	Excavation (whole root system)	Coker (1958), Rogers and Head (1969)	Phlorizin and its breakdown products (Molisch, 1937; Putnam and Weston, 1986; Rice, 1987)
Parthenium argentatum	Plane intersect (bisect wash method)	Muller (1946)	*trans*-Cinnamic acid (Bonner and Galston, 1944; Bonner, 1946)
Pinus taeda	Excavation (whole root system)	Mou *et al.* (1995)	Unknown
Pleuraphis (Hilaria) rigida	Excavation (whole root system)	Nobel (1981), Nobel and Franco (1986)	Unknown
Prunus persica	Excavation (whole root system)	Bini and Chisci (1961)	Prunasin and its breakdown products (Israel *et al.*, 1973; Gur and Cohen, 1989)
Zanthoxylum americanum	Excavation (whole root system)	Reinartz and Popp (1987)	Unknown

of spatial segregation among apple root systems varied greatly among individual trees, whereas Atkinson *et al.* (1976) observed that the degree of root system overlap increased with the density of apple trees. Other agricultural species that may segregate whole-root systems include soybeans (*Glycine max*) (Raper and Barber, 1970) and onions (*Allium cepa*) (Baldwin and Tinker, 1972).

Large-scale root system segregation has also been reported in natural ecosystems. Brisson and Reynolds (1994) reported that whole-root systems of *Larrea tridentata* growing in the Chihuahuan Desert were 4–5 times more segregated than would be predicted by aboveground patterns (Figure 4), and corroborated other reports of low overlap among *Larrea* root systems in other locations (Cannon, 1911; Singh, 1964; Chew and Chew, 1970; Ludwig, 1977). Mou *et al.* (1995) determined that root systems of individual plants of *Liquidambar styraciflua* and *Pinus taeda* overlapped less with conspecifics than predicted by a null model of symmetrical horizontal root distribution (Figure 5). The studies of *Larrea* and *Liquidambar* and *P. taeda* are the only examples in which measured patterns were compared quantitatively against null models. The desert bunchgrass *Pleuraphis (Hilaria) rigida* also shows little horizontal overlap of roots with those of conspecific neighbours (Nobel and Franco, 1986), and the same has been shown for some populations of the semi-desert shrub *Parthenium argentatum* (Bonner and Galston, 1944; Muller, 1946) (Figure 6). Reinartz and Popp (1987) observed virtually no overlap of roots among adjacent clones of northern prickly ash, *Zanthoxylum americanum*. In all of these cases, the degree of root segregation varied among individuals or populations.

Large-scale, whole-root system segregation also occurs between different species. For example, species in the genera *Acacia*, *Casuarina* and *Eucalyptus* appear to exclude roots of certain grasses and forbs from the soil volume occupied by their root systems by mechanisms other than resource competition (Story, 1967; Lange and Reynolds, 1981). Similar phenomena have been reported for *Juniperus osteosperma* in the southwestern United States (Jameson, 1970), *Quercus douglasii* in California (Callaway *et al.*, 1991), and a number of other tree species (Rice, 1984, 1995). Black walnut (*Juglans nigra*) roots exclude the roots of other plants, including alfalfa, tomato, potato and apple (Massey, 1925; Brooks, 1951). Roots of peach trees were excluded from the rooting volumes of *Medicago sativa* and *Trifolium pratense,* but not from those of two grass species (Bergamini, 1965) (Figure 1). Stocker (1928) found little overlap among root systems of several desert species in Egypt.

Vertical root stratification has been reported for shrubs and trees co-occurring with grasses (Cable, 1969; Nye and Tinker, 1977; Callaway *et al.*, 1991) and ericaceous shrubs (Persson, 1980), for mixed forest stands (Mikola *et al.*, 1966; McQueen, 1968; Büttner and Leuschner, 1994), for mixtures of grass-

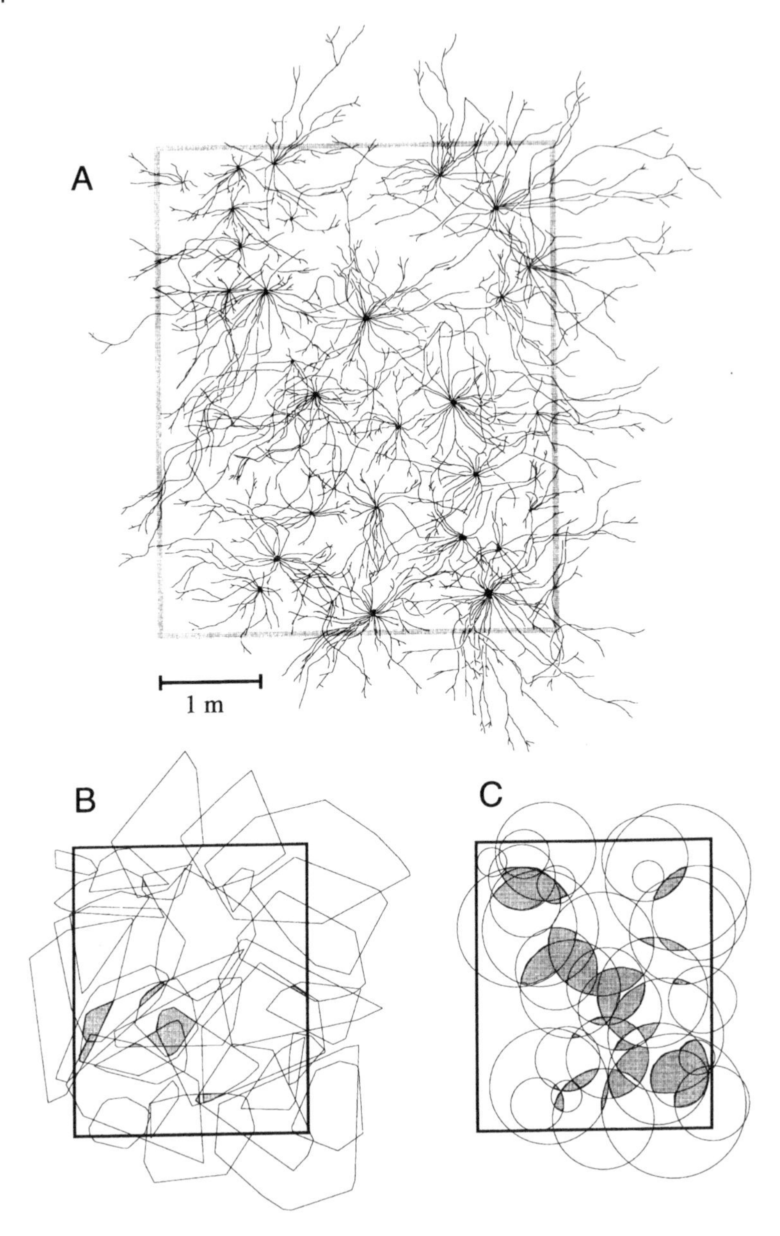

Fig. 4. Spatial segregation of root systems in *Larrea tridentata*. (A) Map of excavated root systems. The rectangle represents the border of the plot. (B) Map of root polygons. (C) Map of hypothetical root systems: circular in shape, centred on the plant location and of surface area equal to their corresponding root polygon. In (B) and (C), the shaded area represents the surface where there is extensive overlap (at least four root areas). Reproduced from Brisson and Reynolds (1994).

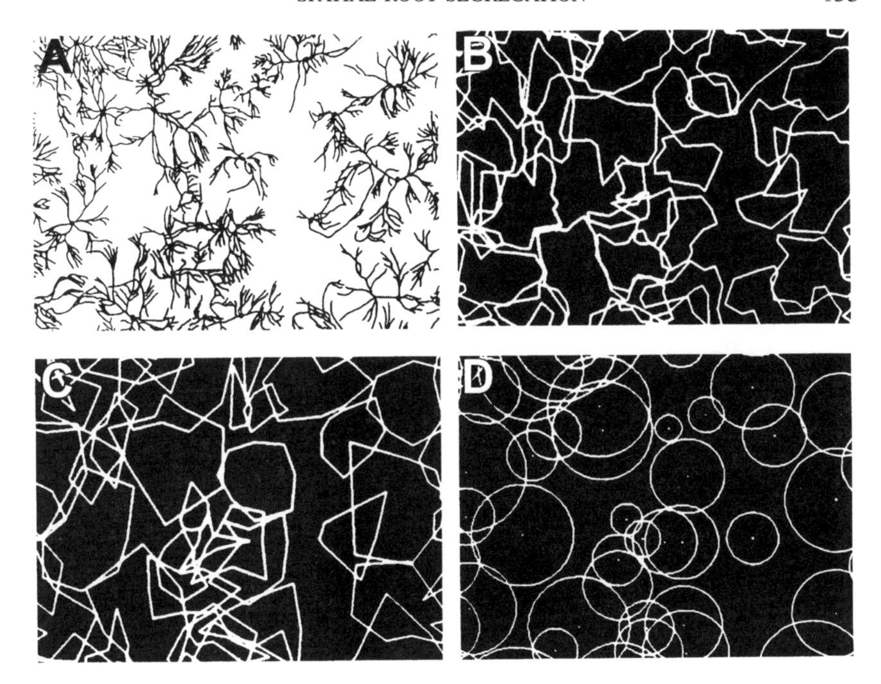

Fig. 5. Overlap patterns of surface roots (0–10 cm) in a stand of sweetgum, *Liquidambar styraciflua*, as determined by excavation. (A) Root map from a photograph. (B) Polygons developed from this photograph. Maps of hypothetical root systems, centred on the plant location and of surface area equal to their corresponding root polygon. (C) Octagons. (D) Circles. The polygons developed from the photo have significantly less overlap than the octagons and circles ($p < 0.005$). Reproduced from Mou *et al.* (1995).

land species (Weaver, 1919, 1920; Parrish and Bazzaz, 1976; Berendse, 1982; Fitter, 1986; Mamolos *et al.*, 1995), for shrub associations (Kummerow *et al.*, 1977; Wright and Mueller-Dombois, 1988; D'Antonio and Mahall, 1991), and for a variety of desert plant associations (Cannon, 1911; Cody, 1986a; Manning and Barbour, 1988; Franco and Nobel, 1990; Briones *et al.*, 1996). The authors of most of these studies considered inherent genetic differences in root architecture or resource depletion to be the main factors causing root system segregation (see Parrish and Bazzaz, 1976; Cody, 1986b).

V. SEGREGATION OF INDIVIDUAL ROOTS

Small-scale spatial segregation among individual roots has been reported in fewer studies than large-scale segregation. Caldwell *et al.* (1991, 1996) demonstrated segregation at very small scales between fine roots of the desert

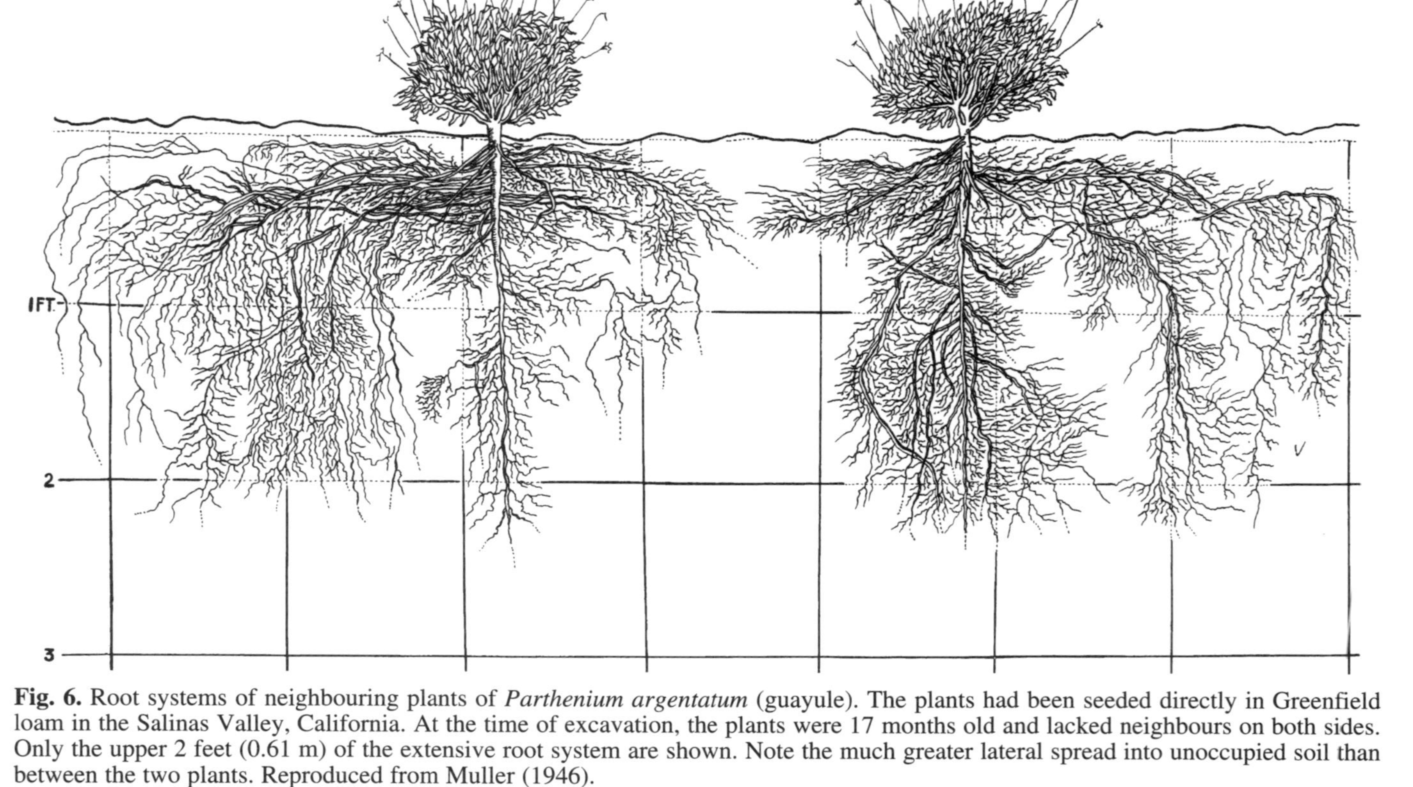

Fig. 6. Root systems of neighbouring plants of *Parthenium argentatum* (guayule). The plants had been seeded directly in Greenfield loam in the Salinas Valley, California. At the time of excavation, the plants were 17 months old and lacked neighbours on both sides. Only the upper 2 feet (0.61 m) of the extensive root system are shown. Note the much greater lateral spread into unoccupied soil than between the two plants. Reproduced from Muller (1946).

shrub *Artemisia tridentata* and those of either one of the grasses *Pseudoroegneria spicata* and *Agropyron desertorum*. They found that nearest-neighbour distances demonstrated clumping when the roots of all species were examined together, but that roots were segregated or "avoid each other" with respect to species (Caldwell *et al.*, 1996). Shrub root density was higher in patches shared with *Agropyron* than in those shared with *Pseudoroegneria* and was correlated with shrub size (Caldwell *et al.*, 1996). In other experiments, Krannitz and Caldwell (1995) found that growth rates of the roots of *P. spicata* decreased after contact with roots of some genotypes of *A. desertorum*. In laboratory studies, roots of the desert shrub *Larrea tridentata* inhibited the growth of other *Larrea* roots and the roots of another desert shrub, *Ambrosia dumosa* (Mahall and Callaway, 1991, 1992). Intraspecific root contact between *Ambrosia dumosa* individuals caused inhibition of growth of contacting roots (Mahall and Callaway, 1991, 1996).

Other studies have documented micro-scale segregation of roots in heterogeneous soil environments, where factors other than root interactions appeared to be the cause for the observed segregation. For example, roots of the grass *Molinia caerulea* and the dwarf shrub *Erica tetralix* co-occurring in a wet heathland varied in abundance in different soil pore-sizes and these distributions correlated with different tolerances to anoxia (Sheik and Rutter, 1969). Roots may also proliferate preferably in litter layers (St John, 1983) or certain soil horizons, or follow canals left by dead roots or the soil fauna, thus producing de facto root segregation without direct root interactions.

VI. MECHANISMS OF ROOT SEGREGATION

A. Indirect Root Interactions

Roots interact indirectly via resource competition. Densities and spatial patterns of the roots of a number of plant species have been shown to be affected by competitors (Bookman and Mack, 1982; Fitter, 1986; Gordon *et al.*, 1989; D'Antonio and Mahall, 1991; Nobel, 1997) and shifts in root distributions of one species are often attributed to resource depletion by the roots of another species. Harper (1985) stated that species of clonal organisms appeared to exhibit regulatory mechanisms to reduce overlap between the same or different individuals, but thought that mechanisms other than competitive plasticity were unlikely. Root segregation may occur when a species exploits soil resources at such a rate or to such a degree that the roots of competitors do not proliferate in the same rooting volume (Bunce *et al.*, 1977; DeLucia and Schlesinger, 1990). Such rapid depletion of resources may directly benefit the plants that capture them and inhibit the growth of neighbour's roots into their root space, but rapid resource uptake is often accompanied by lower resource-use efficiencies (DeLucia and Schlesinger 1990, 1991). Bunce *et al.* (1977) argued that several deciduous trees in the eastern

USA excluded other species from their understories by keeping water resources below the tolerance levels of the subcanopy invaders. *Artemisia tridentata*, a dominant shrub on typical soils of the Great Basin of North America, transpires at much higher rates than *Pinus ponderosa* and *P. jefferyi*, which are restricted to low-pH soils that are not tolerated by *A. tridentata* (DeLucia *et al.*, 1988; DeLucia and Schlesinger, 1990, 1991). Pre-dawn water potentials of *A. tridentata* drop to levels that are 2–3 times below that which the much more water-use-efficient pines can tolerate. DeLucia and Schlesinger (1990, 1991) hypothesised that pine roots could not function in the low-water potential environments created by *A. tridentata*, and later experiments showed that shrubs did competitively exclude pines from normal desert soils (Callaway *et al.*, 1996). Harris (1967) documented rapid pre-emption of soil water resources by the annual grass *Bromus tectorum*. This pre-emption inhibited growth of the perennial grass *Pseudoroegneria spicata* (Harris, 1967). In this case, however, pre-emption was due to temporal separation of root activity (also see Fitter, 1986, 1987) rather than differences in physiological capability. In all of these cases the exclusion of a neighbour's roots by the exhaustive consumption of a particular resource may restrict access by neighbours to other resources that are not as rapidly consumed. However, the roots of a single plant commonly overlap much more than roots of different conspecific individuals. Since all members of a species have the same resource requirements and uptake capabilities, it is difficult to envision how competitive interactions could segregate roots among conspecifics but not roots within an individual.

B. Direct Root Interactions

In addition to responding to resource gradients created by neighbours, roots may also respond to the direct effects of other roots. Mechanisms for direct root interactions include toxins exuded from roots that diffuse into the rhizosphere, non-toxic chemical signals and electrical signals.

1. Allelopathy

The term "allelopathy" was originally used by Hans Molisch in 1937 to describe all chemically mediated stimulatory and inhibitory interactions between plants. Allelopathy is now commonly used to denote chemical inhibition and, together with resource competition, has been proposed as a subcategory of "interference" (Harper, 1961; Muller, 1969). Root-mediated allelopathy and other forms of toxic plant interactions in the soil, often referred to as "soil sickness" (Börner, 1960), have long been described in the literature (e.g. de Candolle, 1832) and evidence for the inhibitory effects of many different root exudates and leachates on plant growth and development has been reviewed elsewhere (e.g. Loehwing, 1937; Grümmer, 1955; Börner, 1960; Woods, 1960; Whittaker, 1970; Whittaker and Feeny, 1971; Rice, 1984; Fischer and Quijano, 1985; Williamson, 1990; Inderjit *et al.*, 1995).

Aboveground plant parts—leaves, stems, flowers, fruits and seeds—can produce substances that can leach into rooting zones and play a role in spatial segregation of roots (Börner, 1960; Rovira, 1969; Nilsson, 1994), but spatial segregation of root systems has primarily been attributed to toxic root exudates (Loehwing, 1937; Grümmer, 1955; Börner, 1960; Woods, 1960; Rice, 1984; Fischer and Quijano, 1985; Inderjit *et al.*, 1995). Most evidence for the effects of root exudates on plant interactions comes from studies of agricultural and horticultural plants (Table 1). In many cases, a specific toxin has been implicated as the putative defence mechanism, but few studies have succeeded in demonstrating that root system segregation is caused by a particular toxin.

Massey (1925) hypothesised that the inhibitory effect of *Juglans nigra* (black walnut) trees on other species, such as alfalfa and tomato, was caused by juglone, a quinone, which is abundant in walnut roots and leaves. Numerous studies have now supported his hypothesis and demonstrated that juglone concentration in the soil under walnut trees is often high enough to inhibit other plants (Brooks, 1951; Fisher, 1978; Ponder and Tadros, 1985). Root segregation among apple trees (Coker, 1958; Rogers and Head, 1969) appears to be caused by a glycoside, phlorizin, a natural constituent of apple root bark, which has been found to inhibit root growth in bioassays (Börner, 1960). In peach, a similar role may be played by breakdown products of a glycoside, prunasin, a constituent of peach root bark (Börner, 1960; Gur and Cohen, 1989). The Australian tree *Grevillea robusta* has been found to produce a water-soluble compound in its living roots that adversely affects and eventually kills conspecific seedlings when their roots come in contact with the roots of mature trees (Webb *et al.*, 1967). In this system, conspecific seedlings cannot root in the soil volume occupied by the roots of adults, but seedlings of other species are not affected. The intraspecific root system segregation observed in the rubber plant guayule, *Parthenium argentatum* (Muller, 1946) (Figure 6), may be due to *trans*-cinnamic acid, which is exuded from the roots (Bonner and Galston, 1944; Bonner, 1946). The presence of freshly cut roots in the soil inhibited growth of guayule plants, but no traces of *trans*-cinnamic acid could be detected in the soil from undisturbed stands of *P. argentatum* (Bonner, 1946).

The effects of root exudates have been experimentally manipulated by adding material with adsorptive capability to soil. In 1907, Schreiner and Reed found that roots of *Lolium perenne* plants would grow away from space occupied by the roots of other individuals. When they added "carbon black" to the substrate, this effect of roots on each other disappeared. Activated carbon, which is probably very similar to the carbon black used by Schreiner and Reed (1907), has little affinity for inorganic electrolytes such as nutrients but strong adsorption capabilities for organic molecules (Cheremisinoff and Ellerbusch, 1978). Mahall and Callaway (1992) found that activated carbon ameliorated the negative effects *Larrea* roots had on each other and on the roots of *Ambrosia dumosa* (see section VII.B). Belliveau and Callaway (1998) found that the roots of *Festuca idahoensis* grew significantly slower as they approached the roots of *Centaurea maculosa*, a noxious invasive weed in North America, but when activated carbon was added to the soil the effects of *Centaurea* were reduced. Activated carbon is not a precise experimental tool for manipulating particular root exudates, and it may have other unwanted effects, but its effects on the interactions among plants (also see Nilsson, 1994) provides evidence for chemically mediated direct interactions among roots.

Direct causal links between specific chemicals exuded from roots, spatial root segregation and the fitness of interacting plants have not been demonstrated (but see section VII). This is due, in part, to the absence of studies designed to make such links, but also to logistic limitations. One of the greatest difficulties is quantifying compounds and simultaneously detecting their source, destination and effects. Further problems arise when compounds are degraded rapidly. For example, phlorizin exuded from apple roots breaks down quickly into non-toxic compounds (Börner, 1971), and polyacetylenes, highly toxic exudates from the roots of *Centaurea* species, are volatile and only remain in the soil for short times (Bohlmann *et al.*, 1973; Stevens, 1986). Microbial alteration of root exudates may inactivate allelopathic root exudates, but they may also convert harmless exudates into active ones. For example, the microbial breakdown of prunasin into toxic HCN and benzaldehyde was found to occur in the rhizosphere of actively growing peach trees (Gur and Cohen, 1989).

Segregation among individual fine roots of *Artemisia tridentata*, *Pseudoroegneria spicata* and *Agropyron desertorum* (Caldwell *et al.*, 1991, 1996; section V) may have been caused by direct interactions among roots. There was no indication that the "species-specific interactions were the result of resource competition, since there were no obvious patterns between the proportions of shrub and grass roots of the two species combinations with microsite nutrient concentrations" (Caldwell *et al.*, 1996). Caldwell *et al.* (1996) held that their evidence strongly suggested that a discriminating recognition mechanism was involved. Huber-Sannwald *et al.* (1996, 1997) reported corroborating evidence for species-specific recognition among *Agropyron* and *Pseudoroegneria*, as well as *Elymus lanceolatus*. However, the putative recognition mechanism between these species may have been either allelopathic or a non-toxic signal.

2. *Non-Toxic Signals*

It has long been hypothesised that non-toxic, non-allelopathic signals between roots occur (Lund, 1947; Cohen, 1970), and evidence has accumulated for the effects of chemical signals ("infochemicals"; Dicke and Sabelis, 1988) on interactions among plants and between roots (Aphalo and Ballaré, 1995; Bruin *et al.*, 1995). Signalling mechanisms affect the ability of the roots of parasitic plants to move towards the roots of hosts (Lynn and Boone, 1993), and it is reasonable to speculate that similar signals may allow the roots of autotrophic plants to detect those of potential competitors. Chemical signals play a role in the interaction between roots and bacteria (e.g. *Rhizobium*, *Agrobacterium*), fungi and parasitic plants (Smith and Van Staden, 1995). Signal molecules include phenolic compounds, chinones, flavonoids, oligosaccharides and lectins (Callow, 1984; Gianinazzi-Pearson, 1984; Lynn and Chang, 1990; Goethals *et al.*, 1993; Kapulnik *et al.*, 1993; Lynn and Boone, 1993).

Electrical signals may also mediate communications between roots (Lund, 1947). Roots generate ionic currents (Miller and Gow, 1989a,b), and they are also capable of sensing electric fields and pH gradients, as evidenced by their curving towards anodes or cathodes depending on field strength (Schrank, 1959; Marcum and Moore, 1990; Stenz and Weisenseel, 1991, 1993). Zoospores of the plant pathogen *Phytophthora palmivora* (Oomycetes) are sensitive to electric fields and appear to target their host roots using a combination of chemotaxis and electrotaxis (Gow *et al.*, 1992; Morris *et al.*, 1992).

Distinguishing between the effects of toxins and those of non-toxic signals is difficult. Mahall and Callaway (1991, 1992, 1996) reported that the roots of a desert shrub, *Ambrosia dumosa*, responded to genotype-specific, contact-mediated signals that were not generally toxic. In these studies, roots that contacted roots of a separate, conspecific plant from the same population or of the same genotype exhibited strong growth inhibition. In the same experiments, roots that did not contact those of other plants continued to grow. When contact occurred between sister roots on the same plant or between roots of plants from different populations, no inhibition occurred. Furthermore, when activated carbon was added to soil to adsorb root exudates, it had no effect on the root interactions among *Ambrosia* plants. Another case for root communication has been reported for roots of the desert grass *Pseudoroegneria spicata* encountering roots of different genotypes of another grass species, *Agropyron desertorum* (Krannitz and Caldwell, 1995). For both species, root segregation occurred under conditions of high resource availability, indicating that resource depletion was an unlikely explanation (Caldwell *et al.*, 1996), and that segregation involved self/non-self-recognition signals that were not generally toxic.

VII. PLANT TERRITORIES

Territoriality is defined as the defence and exclusive use of space (Maher and Lott, 1995). The ubiquity of territorial behaviour among animal taxa has established territoriality as a central and unifying concept in fields ranging from ecology and sociobiology to human geography (Wilson, 1975; Malmberg, 1980; Krebs and Davies, 1981). Species of insects, crustaceans, fishes, amphibians, reptiles, birds, mammals and fungi have been found to establish territories by defending resources such as food and habitat against conspecific and interspecific neighbours. Active root segregation suggests that territoriality may extend to plants.

We use the term "territory" in order to emphasise the potential for conceptual parallels and contrasts between some kinds of root segregation and animal territories. By definition, territories must be defended. Therefore de facto root segregation due to soil structure, root architecture or physiological constraints that have been discussed elsewhere does not constitute territoriality. For animals, defences may vary from aggressive exclusion of intruders to mere advertisement by chemical or other signals (Anderson and Hill, 1965; Wilson, 1975), and there are many defence mechanisms including visual displays, vocalisations, release of toxins and lethal combat (Wilson, 1975; Malmberg, 1980). As in the animal kingdom, territorial defences in plants can often only be inferred from their effects. For example, bird vocalisations do not always have unique properties that mark them as either sexual advertisement, territorial defence or warning calls. Such "functions" can only be deduced from the observed reactions of other animals. Similarly, the defence mechanisms of root-mediated allelopathy and non-toxic signals must be causally linked to spatial segregation of roots and root systems and increased resource acquisition to constitute territoriality. These links may prove to be elusive since indisputable evidence for such links has seldom been found for any organism. On the other hand, plants may provide superior systems for experimental manipulation and modelling of territoriality and, if included in more studies of territoriality, invigorate empirical and theoretical progress.

A. Defence of Space

The space occupied by a plant's canopy or root system is often used as an index of its ability to acquire resources (Bella, 1971; Czárán and Bartha, 1992). It is well known that plants generally grow slower when space is physically restricted (e.g. "pot-binding") even when other resources are abundant (Richards and Rowe, 1977; Tschaplinski and Blake, 1985; Wilson and Newman, 1987; Gurevitch *et al.*, 1990). Photosynthetic capacity may also be adversely affected by small rooting volumes (Thomas and Strain, 1991). Space acquired by individual plants in monocultures, measured as Voronoi polygons constructed around individual plants, has been found to be positively correlated with growth and survival (Mithen *et al.*, 1984). Moreover, plants appear to be able to compete for space independently of nutrient, water or light resources. McConnaughay and Bazzaz (1991) examined the effects of different levels of belowground space on several annual species and found that, when equal amounts of nutrients were provided, plants grown in a greater volume of soil increased growth and reproductive output. Fragmentation of belowground space by plastic, artificial root systems also had a negative effect on plant growth when resource levels were low (McConnaughay and Bazzaz, 1992). Resource-independent space requirements are likely to be connected to morphological constraints imposed by root system architecture. Root systems of many plant species may not possess the plasticity necessary to utilise resources efficiently in a small or fragmented environment (Parrish and Bazzaz, 1976; McConnaughay and Bazzaz, 1992; Fitter, 1994). The likelihood that space is itself a limiting resource for plants, suggests that allelopathy and signal-driven root segregation may generate territoriality regardless of their effects on the acquisition of other resources.

B. *Larrea tridentata*: A Case Study

The most comprehensive case for territoriality in plants is that for *Larrea tridentata* (creosote bush), a dominant shrub in much of the North and South American hot desert. First, the roots of *Larrea* appear to possess a mechanism for the defence of space and show intra- and interspecific root segregation at small scales (millimetres to centimetres). Mahall and Callaway (1991, 1992) demonstrated that root growth of other conspecifics and another commonly associated species, *Ambrosia dumosa*, significantly decreased when approaching or contacting target *Larrea* roots. These negative effects were significantly ameliorated when activated carbon, an effective adsorbent of organic compounds, was added to the soil. Secondly, excavation of whole root systems of *Larrea* show them to be highly segregated from each other (Figure 4; Cannon, 1911; Singh, 1964; Chew and Chew, 1970; Ludwig, 1977; Brisson and Reynolds, 1994). Hutto *et al.* (1986) and McAuliffe (1988) also showed that *Larrea* was spatially disassociated with most other perennial plants in Mojave and Sonoran Desert communities. Finally, field experiments conducted by Fonteyn and Mahall (1981) showed that the removal of *Ambrosia dumosa* from around *Larrea* shrubs increased their pre-dawn water potentials. Although various pieces of the territoriality puzzle exist for many other species, to our knowledge, this is the only system in which links can be made between a mechanism for the defence of space, field evidence for small- and large-scale root segregation, and evidence that more space may translate into acquiring more of another limiting resource.

C. Territoriality and Competition

The presumed advantage of defending space, and the resources contained within it, is that it is less costly than direct competition for the same resources in shared space (Wilson, 1975). In other words, "scramble" competition, where the organisms interact directly with resources and indirectly with each other, is thought to be less energetically favourable than "contest" competition (Crombie, 1947), where the organisms interact directly with their competitors (Nicholson, 1954; McIntosh, 1992; Gordon and Kulig, 1996). The extent of root-system overlap is considered a good measure of the potential for resource competition (analogous to scramble competition) among roots (Caldwell and Richards, 1986; Morris and Myerscough, 1991). Therefore measuring the relationship between the extent of root overlap and plant performance may provide insight into the costs and benefits of defending space. The potential benefits of preferentially utilising non-occupied space was well illustrated by Brisson and Reynolds (1997) who modelled the effects of "compensatory" root growth (the ability of roots to grow disproportionally into and occupy space that is free of neighbour's roots) on population responses. Their simulations showed that plants that maintained segregated root systems were likely to be better competitors than plants with overlapping root systems, and that the former were able to more completely occupy available space, grow larger than non-compensatory plants and, in initially mixed populations, eventually eliminate non-compensatory plants. Of import to the defence of space by plants was that simulated populations of plants with segregated root systems experienced far less density-dependent mortality than those with overlapping roots.

Root system segregation has primarily been documented among conspecific neighbours (Muller, 1946; Coker, 1958; Israel *et al.*, 1973; Nye and Tinker, 1977; Nobel and Franco, 1986; Reinartz and Popp, 1987; Brisson and Reynolds, 1994; Mou *et al.*, 1995), suggesting that avoiding scramble competition may provide particular advantages over direct competition for resources among related plants, especially among genetically identical clones (Schenk, 1998). The genetic relationship among neighbours should shift the balance of costs and benefits of territoriality (Mappes *et al.*, 1995; Schenk, 1998), and, for animals, territorial behaviour and effects are often different among unrelated conspecific neighbours than among kin (Francis, 1973; Kawata, 1987; Lambin and Krebs, 1993; Mappes *et al.*, 1995; but see Brown and Brown, 1993a,b). Roots of the desert shrub *Ambrosia dumosa* have much stronger recognition responses to the roots of other genetically identical individuals (clones) and plants from the same population than to plants from distant populations (Mahall and Callaway, 1996). In root observation chambers, the roots of individuals from distant populations became intermixed, whereas the roots of individuals remained segregated, suggesting that direct competition would be much greater between the former. Roots of the same individual clone of

Zanthoxylum americanum crossed each other frequently and intermixed, but roots of different individuals rarely did (Reinartz and Popp, 1987). Reduced resource competition among closely related plants has been found in *Phytolacca americana* (Willson *et al.*, 1987) and *Plantago lanceolata* (Tonsor, 1989), but root segregation was not measured.

The only way for a plant to outcompete its neighbour directly for a resource is to take it before the neighbour can; a tactic that does not promote resource-use efficiency. If plants defend resources rather than compete directly for them, only then can they afford to use them efficiently. For example, conservative water use in arid environments would not be an advantage for plants with roots that intermingled with those of their competitors, unless water was stored internally. Efficient water-use would only be an advantage to species without profligate neighbours (Cohen, 1970; Fischer and Turner, 1978; Jones, 1993; Aphalo and Ballaré, 1995).

Trees and shrubs that have deep roots, and presumably avoid intense competition for water in shallow soils, have been found to have higher water-use efficiencies than other co-occurring species with shallow roots (Ehleringer and Cooper, 1988; DeLucia and Schlesinger, 1990). In cold deserts, pines with high water-use efficiencies and low rates of water use were competitively excluded by shrubs that transpired more rapidly and did use water as efficiently (Callaway *et al.*, 1996). In contrast, *Larrea tridentata* is a relatively shallow-rooted species, yet was the most water-use-efficient desert shrub in a study of 22 co-occurring species (Ehleringer and Cooper, 1988). As described above, *Larrea* roots have strong negative effects on the roots of other species and its root systems are highly segregated from those of others (Figure 2; Mahall and Callaway, 1991, 1992; Brisson and Reynolds, 1994). In dry woodlands of Australia, *Acacia*, *Casuarina* and *Eucalyptus* species exclude herbaceous species from their rooting zone (Story, 1967). All of the tree species used water more slowly than the herbaceous species.

D. Density, Resource Availability and Territoriality

The size and effectiveness of territories varies with neighbour density and resource availability (Dill, 1978; Ebersole, 1980; Hixon, 1980; Provencher and Vickery, 1988; Brown and Brown, 1993a). The desert grass *Pleuraphis (Hilaria) rigida* has been shown to have little root overlap with neighbouring conspecifics (Nobel, 1981). Removal of conspecific neighbours (reducing density) improved water relations of target plants (Robberecht *et al.*, 1983) and, over a period of several years, the remaining plants expanded their root systems into the soil volumes formerly occupied by neighbours (Nobel and Franco, 1986). The root systems of apple trees are often highly segregated (Coker, 1958) but territoriality appears to break down at higher tree densities as root system overlap increases (Atkinson *et al.*, 1976).

Since territoriality often degenerates when resources are readily available, the relative importance of territoriality in plant communities may vary among environments. Territoriality may not be as important in mesic, nutrient-rich communities as direct competition for resources. Fitter (1987) concluded that there was little evidence for spatial segregation of roots; however, studies reviewed by Fitter were conducted primarily in mesic environments. In contrast, Cody (1986b) argued that belowground root segregation was the rule rather than the exception, but restricted his comparison of root architecture among woody perennials to arid environments. Temperature and soil moisture fluctuations influence root exudation (Curl and Truelove, 1986; Bar-Yosef, 1991; Tang *et al.*, 1995), and plants in nutrient- or water-poor sites often produce higher concentrations of secondary metabolites than conspecifics growing in fertile or well-watered soil (McKey *et al.*, 1978; Gershenzon, 1979; Rosenthal and Janzen, 1979; Mihaliak *et al.*, 1987; Northup *et al.*, 1995). Stressful conditions also promote root exudation in general (Koeppe *et al.*, 1976; Curl and Truelove, 1986), and often cause increased allocation to total root biomass and fine root production (Callaway, 1990). Hall *et al.* (1982, 1983) found that phenolic concentrations in *Helianthus annuus* tissues increased under nutrient stress and *H. annuus* litter produced under nutrient stress had stronger allelopathic effects on *Amaranthus retroflexus* seedlings. Tang *et al.* (1995) found that the inhibitory effects of the methanol fraction of *Cyperus rotundus* (purple nutsedge) root exudates on the radicle elongation of lettuce were 90% greater when *C. rotundus* was grown under water stress. These results suggest that the intensity of mechanisms that can lead to spatial root segregation increases when plants are subjected to limited resources.

Stress induced by herbivory may also increase allocation to the defence of belowground space. Secondary chemicals involved in defence against pathogens and herbivores, and inducible by herbivory, can also be allelopathic. For example, gramine and hordenine, chemicals produced by *Hordeum* species, can inhibit fungal and animal consumers and also suppress other plant species (Lovett and Hoult, 1995). Species in the Brassicaceae family are defended from consumers by glucosinolates and myrosinase (Bones and Rossiter, 1996), and these defence compounds can have allelopathic effects on plant neighbours (Oleszek, 1987; Garner *et al.*, 1996). Schmitt *et al.* (1995) found that diterpene alcohols produced by the brown seaweed *Dictyota menstrualis* deterred feeding by different species of herbivores and prevented the growth of a bryozoan that overgrew the surface of the seaweed. The roots of *Centaurea maculosa*, an invasive weed in North America, limit root growth and total biomass of native *Festuca idahoensis* via root exudates (Belliveau and Callaway, 1998). *Centaurea* plants that had been subjected to moderate herbivory from cabbage loopers had even stronger effects on *F. idahoensis*, and the roots of grazed *Centaurea* plants exuded greater levels of some carbohydrates than plants protected from herbivory (Callaway *et al.*, 1998). *Centaurea maculosa* produces the chemical cnicin and numerous polyacetylenes, which have both anti-herbivore and allelopathic capabilities (Fletcher and Renny, 1963; Bohlmann *et al.*, 1973; Stevens, 1986; Kelsey and Locken, 1987; Landau *et al.*, 1994).

VIII. TERRITORIALITY AND THE REGULATION OF PLANT POPULATIONS

Whether or not territoriality regulates the population of any organism is controversial (Wynne-Edwards, 1962; Williams, 1966; Wilson, 1975) and two competing hypotheses have been proposed. The first, the individual-selection hypothesis, assumes that resources limit populations and territoriality is a mechanism for defending resources when they are in short supply (Lack, 1954; Wilson, 1975). According to this hypothesis, resource-based territoriality should have minor effects on population density (Williams, 1966; Wilson, 1975), or population productivity and fecundity, which are better measures for plants (Harper, 1960).

The "group-selection" hypothesis of territoriality, postulates that territoriality keeps populations below the carrying capacity that would be determined by resources alone (Wynne-Edwards, 1962, 1986). Such regulation would ensure that resources were not spread too thin among competitors and dampen oscillations in productivity as resource availability varied (Wilson, 1980). Went (1970) used a similar argument when he hypothesised that allelopathy may play a role in regulating plant population density below the environmental carrying capacity. This scenario may be even more complex in plant populations because carrying capacities are complex, and the densities and structures of plant populations, which may occupy the same space for hundreds of years, may be determined in exceptionally bad years. Proponents of the individual-selection hypothesis have speculated that regulation of density (or productivity) below the resource-determined carrying capacity implies that territory owners defend resources that they do not actually use, which would not be to their individual advantage, and that "cheater" genotypes would soon disrupt the territorial arrangement (Williams, 1966; Wilson, 1975). Therefore, it has been reasoned that such a system could not evolve by individual selection, and group or kin selection must be invoked to evolve a system of safeguarding resources (Wilson, 1975; Nakamura, 1980; Wynne-Edwards, 1986). However, kin selection and other forms of hierarchical selection (Damuth and Heisler, 1988; Stevens *et al.*, 1995) may function in populations where individuals interact in relatively stable groups, a condition that applies to most plants (Nakamura, 1980; Wilson, D.S., 1980; Wilson, J.B., 1987) and to sessile organisms in general. Recent studies have found evidence for the importance of selection at higher levels than that of the individual in plants (Goodnight, 1985; Stevens *et al.*, 1995) and other observations also suggest that kin-selected adaptations may exist in plants (Nakamura, 1980; Willson *et al.*, 1987; Wilson, 1987; Mahall and Callaway, 1996). Comparing the relationship between territoriality and population regulation of plants with that of animals may shed new light on these competing hypotheses and the debate on the evolution of territoriality and its role in population regulation.

The inchoate evidence for the defence of space and root segregation among plants suggests a level of ecological generality that is rare in a science marked by species- and system-specific idiosyncrasy. Plants may provide an opportunity for manipulative experiments with territories, not possible with mobile animals, and so have potential for tests of long-standing theoretical questions about territoriality.

ACKNOWLEDGMENTS

The authors thank: Alastair Fitter, Robert B. Jackson, Joshua P. Schimel, Stephen I. Rothstein and an anonymous reviewer for their helpful comments on the manuscript; Bruce H. Tiffney and J. Robert Haller for comments on early drafts; and Jason G. Hamilton and Claus Holzapfel for helpful discussions. We also gratefully acknowledge financial support of this work by a grant from the Andrew W. Mellon Foundation to B.E. Mahall. This paper is dedicated to the memory of Cornelius H. Muller for his pioneering efforts in studies of root distributions and plant interactions.

REFERENCES

Anderson, P.K. and Hill, J.L. (1965). *Mus musculus*: Experimental induction of territory formation. *Science* **148**, 1753–1755.

Aphalo, P.J. and Ballaré, C.L. (1995). On the importance of information-acquiring systems in plant–plant interactions. *Funct. Ecol.* **9**, 5–14.

Atkinson, D., Naylor, D. and Coldrick, G.A. (1976). The effect of tree spacing on the apple root system. *Hort. Res.* **16**, 89–105.

Baldwin, J.P. and Tinker, P.B. (1972). A method for estimating the lengths and spatial patterns of two interpenetrating root systems. *Plant Soil* **37**, 209–213.

Baldwin, J.P., Tinker, P.B. and Marriott, F.H.C. (1971). The measurement of length and distribution of onion roots in the field and in the laboratory. *J. appl. Ecol.* **8**, 543–554.

Baldwin, J.P., Tinker, P.B. and Nye, P.H. (1972). Uptake of solutes by multiple root systems from soil. II. The theoretical effects of rooting density and pattern on uptake of nutrients from soil. *Plant Soil* **36**, 693–708.

Bargioni, G. (1962). Alcune osservazioni sul sistema radicale del diospiro. [Some observations on the root system of persimmon. In Italian with English summary.] *Riv. Ortoflorofrutt. Ital.* **46**, 569–579.

Bargioni, G. (1968). Antagonism among root systems of fruit trees (abstract). In: *Methods of Productivity Studies in Root Systems and Rhizosphere Organisms. International Symposium USSR, August 28–September 12, 1968* (Ed. by M.S. Ghilarov, V.A. Kovda, L.N. Novichkova-Ivanova, L.E. Rodin and V.M. Sveshnikova), pp. 3–4. Nauka, Leningrad.

Barth, J. (1995). Acceptance criteria for using individual-based models to make management decisions. *Ecol. Applic.* **5**, 411–420.

Bartholomew, B. (1970). Bare zone between California shrub and grassland communities: The role of animals. *Science* **170**, 1210–1212.

Bar-Yosef, B. (1991). Root excretions and their environmental effects: Influence on availability of phosphorus. In: *Plant Roots: The Hidden Half* (Ed. by Y. Waisel, A. Eshel and U. Kafkafi), pp. 529–557. Marcel Dekker, New York.

Beck, E., Fusseder, A. and Kraus, M. (1989). The maize root system *in situ*: Evaluation of structure and capability of utilization of phytate and inorganic soil phosphates. *Z. Pflanzenernähr. Bodenk.* **152**, 159–167.

Bella, I.E. (1971). A new competition model for individual trees. *Forest Sci.* **17**, 364–372.

Belliveau, R. and Callaway, R.M. (1998). Relative effects of root competition and allelopathy between *Centaurea maculasa* and a native bunchgrass. *Oecologia* (in review).

Berendse, F. (1982). Competition between plant populations with different rooting depths III. Field experiments. *Oecologia* **53**, 50–55.

Bergamini, A. (1965). Influenza di alcune specie erbacee sulla distribuzione delle radici del pesco. [The influence of some herbaceous species on the distribution of peach roots. In Italian with English summary.] In: *Atti del Congresso del Pesco*, pp. 507–511. Camera di Commercio Industria e Agricoltura di Verona, Verona.

Berlandier, J.L. (1980). Journey to Mexico during the years 1826–1834. (Translated from the French by S.M. Ohlendorf, J.M. Bigelow, and M.M. Standifer. Botanical notes by C.H. Muller and K.K. Muller.) The Texas State Historical Association, Austin, Texas.

Bini, G. and Chisci, P. (1961). Alcune osservazioni sul reciproco comportamento delle radici del pesco e del pero. [Some observations on the effects of roots of peach and pear. In Italian with English summary.] *Riv. Ortoflorofrutt. Ital.*, 345–352.

Böhm, W. (1979). *Methods of Studying Root Systems*. Springer Verlag, Berlin.

Bohlmann, R., Burkhardt, T. and Zdero, C. (1973). *Naturally Occurring Acetylenes*. Academic Press, London.

Bones, A.M. and Rossiter, J.T. (1996). The myrosinase–glucosinolate system, its organisation and biochemistry. *Physiol. Plant.* **97**, 194–208.

Bonner, J. (1946). Further investigations of toxic substances which arise from guayule plants: Relation of toxic substances to the growth of guayule in soil. *Bot. Gaz.* **107**, 185–198.

Bonner, J. and Galston, A.W. (1944). Toxic substances from the culture media of guayule which may inhibit growth. *Bot. Gaz.* **106**, 185–198.

Bookman, P.A. and Mack, R.N. (1982). Root interaction between *Bromus tectorum* and *Poa pratensis*: A three-dimensional model. *Ecology* **63**, 640–646.

Börner, H. (1960). Liberation of organic substances from higher plants and their role in the soil sickness problem. *Bot. Rev.* **26**, 393–424.

Börner, H. (1971). German research on allelopathy. In: *Biochemical Interactions among Plants*, pp. 52–55. National Academy of Sciences, Washington, DC.

Briones, O., Montaña, C. and Ezcurra, E. (1996). Competition between three Chihuahuan desert species: Evidence from plant size-distance relations and root distribution. *J. Veg. Sci.* **7**, 453–460.

Brisson, J. and Reynolds, J.F. (1994). The effect of neighbors on root distribution in a creosotebush (*Larrea tridentata*) population. *Ecology* **75**, 1693–1702.

Brisson, J. and Reynolds, J.F. (1997). Effects of compensatory growth on population processes: A simulation study. *Ecology* **75**, 1693–1702.

Brooks, M.G. (1951). Effects of black walnut trees and their products on other vegetation. *West Virginia Univ. Agric. Exp. Station Bull.* **347**, 1–31.

Brown, G.E. and Brown, J.A. (1993a). Do kin always make better neighbours? The effects of territory quality. *Behav. Ecol. Sociobiol.* **33**, 225–231.

Brown, G.E. and Brown, J.A. (1993b). Social dynamics in salmonid fishes: Do kin make better neighbors? *Anim. Behav.* **45**, 863–871.

Bruin, J., Sabelis, M.W. and Dicke, M. (1995). Do plants tap SOS signals from their infested neighbors? *Trends Ecol. Evol.* **10**, 167–170.

Bunce, J.A., Miller, L.N. and Chabot, B.F. (1977). Competitive exploitation of soil water by five eastern North American tree species. *Bot. Gaz.* **138**, 168–173.

Büttner, V. and Leuschner, C. (1994). Spatial and temporal patterns of fine-root abundance in a mixed oak–beech forest. *For. Ecol. Manag.* **70**, 11–21.

Cable, D.R. (1969). Competition in the semidesert grass-shrub type as influenced by root systems, growth habits, and soil moisture extraction. *Ecology* **50**, 27–38.

Caldwell, M.M. and Richards, J.H. (1986). Competing root systems: Morphology and models of absorption. In: *On the Economy of Plant Form and Function* (Ed. by T.J. Givnish), pp. 251–273. Cambridge University Press, Cambridge.

Caldwell, M.M., Manwaring, J.H. and Durham, S.L. (1991). The microscale distribution of neighbouring plant roots in fertile soil microsites. *Funct. Ecol.* **5**, 765–772.

Caldwell, M.M., Manwaring, J.H. and Durham, S.L. (1996). Species interactions at the level of fine roots in the field: Influence of soil nutrient heterogeneity and plant size. *Oecologia* **106**, 440–447.

Caldwell, M.M., Richards, J.H., Manwaring, J.H. and Eissenstat, D.M. (1987). Rapid shifts in phosphate acquisition show direct competition between neighbouring plants. *Nature* **327**, 615–616.

Callaway, R.M. (1990). Effects of soil water distribution on the lateral root development of three species of California oaks. *Amer. J. Bot.* **77**, 1469–1475.

Callaway, R.M., DeLuca, T. and Belliveau, W. (1998). Herbivores used for biological control increase the competitive ability of target weeds. *Ecology* (in press).

Callaway, R.M., Nadkarni, N.M. and Mahall, B.E. (1991). Facilitation and interference of *Quercus douglasii* on understory productivity in central California. *Ecology* **72**, 1484–1499.

Callaway, R.M., DeLucia, E.H., Moore, D., Nowak, R. and Schlesinger, W.H. (1996). Competition and facilitation: Contrasting effects of *Artemisia tridentata* on desert vs. montane pines. *Ecology* **77**, 2130–2141.

Callow, J.A. (1984). Cellular and molecular recognition between higher plants and fungal pathogens. In: *Cellular Interactions* (Ed. by H. F. Linskens and J. Heslop-Harrison), pp. 212–237. Springer Verlag, Berlin.

de Candolle, A.P. (1832). *Physiologie Végétale III*. Bechet Jeune, Paris.

Cannon, W.A. (1911). *The Root Habits of Desert Plants*. Carnegie Institution of Washington, Washington, DC.

Cannon, W.A. (1913). Notes on root variation in some desert plants. *Plant World* **16**, 323–341.
Casper, B.B. and Jackson, R.B. (1997). Plant competition underground. *Annu. Rev. Ecol. Syst.* **28**, 545–570.
Cheremisinoff, P.N. and Ellerbusch, F. (1978). *Carbon Adsorption Handbook*. Ann Arbor Science Publishers, Ann Arbor, MI.
Chew, R.M. and Chew, A.E. (1970). Energy relationships of the mammals of the desert scrub (*Larrea tridentata*) community. *Ecol. Monogr.* **40**, 1–21.
Cody, M.L. (1986a). Spacing in Mojave Desert plant communities. II. Plant size and distance relationships. *Isr. J. Bot.* **35**, 109–120.
Cody, M.L. (1986b). Structural niches in plant communities. In: *Community Ecology* (Ed. by J. Diamond and T.J. Case), pp. 381–405. Harper & Row Publishers, New York.
Cohen, D. (1970). The expected efficiency of water utilization in plants under different competition and selection regimes. *Isr. J. Bot.* **19**, 50–54.
Coker, E.G. (1958). Root studies XII. Root systems of apple on Malling rootstocks on five soil series. *J. Hort. Sci.* **33**, 71–79.
Coutts, M.P. (1989). Factors affecting the direction of growth of tree roots. *Ann. Sci. Forest.* **46** (Suppl.), 277s–287s.
Crombie, A.C. (1947). Interspecific competition. *J. Anim. Ecol.* **16**, 44–73.
Curl, E.A. and Truelove, B. (1986). *The Rhizosphere*. Springer Verlag, Berlin.
Czárán, T. and Bartha, S. (1992). Spatiotemporal dynamic models of plant populations and communities. *Trends Ecol. Evol.* **7**, 35–69.
Damuth, J. and Heisler, I.L. (1988). Alternative formulations of multilevel selection. *Biol. Philosoph.* **3**, 407–430.
D'Antonio, C.M. and Mahall, B.E. (1991). Root profiles and competition between the invasive, exotic perennial, *Carpobrotus edulis*, and two native shrub species in California coastal scrub. *Amer. J. Bot.* **78**, 885–894.
DeLucia, E.H. and Schlesinger, W.H. (1990). Ecophysiology of Great Basin and Sierra Nevada vegetation on contrasting soils. In: *Plant Biology of the Basin and Range* (Ed. by C.B. Osmond, L.F. Pitelka and G.M. Hidy), pp. 143–178. Springer Verlag, Berlin.
DeLucia, E.H. and Schlesinger, W.H. (1991). Resource-use efficiency and drought tolerance in adjacent Great Basin and Sierran plants. *Ecology* **72**, 1533–1543.
DeLucia, E.H., Schlesinger, W.H. and Billings, D. (1988). Water relations and the maintenance of Sierran conifers on hydrothermally altered rock. *Ecology* **69**, 303–311.
Dicke, M. and Sabelis, M.W. (1988). Infochemical terminology: Based on cost–benefit analysis rather than origin of compounds? *Funct. Ecol.* **2**, 131–139.
Dill, L.M. (1978). An energy-based model of optimal feeding-territory size. *Theor. Popul. Biol.* **14**, 396–429.
Ebersole, J.P. (1980). Food density and territory size: An alternative model and a test on the reef fish *Eupomacentrus leucostictus*. *Am. Nat.* **115**, 492–509.
Ehleringer, J.R. and Cooper, T.A. (1988). Correlations between carbon isotope ratio and microhabitat in desert plants. *Oecologia* **76**, 562–566.
Escamilla, J.A., Comerford, N.B. and Neary, D.G. (1991). Spatial pattern of slash pine roots and its effect on nutrient uptake. *Soil Sci. Soc. Am. J.* **55**, 1716–1722.
Firbank, L.G. and Watkinson, A.R. 1985. A model of interference within plant monocultures. *J. theor. Biol.* **116**, 291–311.

Fischer, N.H. and Quijano, L. (1985). Allelopathic agents from common weeds. In: *The Chemistry of Allelopathy* (Ed. by A.C. Thompson), pp. 133–147. American Chemical Society, Washington, DC.

Fischer, R.A. and Turner, N.C. (1978). Plant productivity in the arid and semiarid zones. *Ann. Rev. Plant Physiol.* **29**, 277–317.

Fisher, R.F. (1978). Juglone inhibits pine growth under certain moisture regimes. *Soil Sci. Soc. Am. J.* **42**, 801–803.

Fitter, A.H. (1986). Spatial and temporal patterns of root activity in a species-rich alluvial grassland. *Oecologia* **69**, 594–599.

Fitter, A.H. (1987). Spatial and temporal separation of activity in plants communities: Prerequisite or consequence of coexistence. In: *Organization of Communities: Past and Present; 27th Symposium of the British Ecological Society, Aberystwyth, Wales, UK, 1986* (Ed. by J.H.R. Gee and P.S. Giller), pp. 119–139. Blackwell Scientific Publications, Palo Alto, CA.

Fitter, A.H. (1994). Architecture and biomass allocation as components of the plastic response of root systems to soil heterogeneity. In: *Exploitation of Environmental Heterogeneity by Plants: Ecophysiological Processes Above- and Belowground* (Ed. by M.M. Caldwell and R.W. Pearcy), pp. 305–323. Academic Press, San Diego.

Fitter, A.H. and Stickland, T.R. (1991). Architectural analysis of plant root systems 2. Influence of nutrient supply on architecture in contrasting plant species. *New Phytol.* **118**, 383–389.

Fitter, A.H. and Stickland, T.R. (1992). Architectural analysis of plant root systems III. Studies on plants under field conditions. *New Phytol.* **121**, 243–248.

Fitter, A.H., Stickland, T.R., Harvey, M.L. and Wilson, G.W. (1991). Architectural analysis of plant root systems 1. Architectural correlates of root exploitation efficiency. *New Phytol.* **118**, 375–382.

Fletcher, R.A. and Renny, A.J. (1963). A growth inhibitor found in *Centaurea* spp. *Can. J. Plant Sci.* **43**, 475–481.

Fonteyn, P.J. and Mahall, B.E. (1981). An experimental analysis of structure in a desert plant community. *J. Ecol.* **69**, 883–896.

Francis, L. (1973). Clone specific segregation in the sea anemone *Anthopleura elegantissima. Biol. Bull.* **144**, 64–72.

Franco, A.C. and Nobel, P.S. (1990). Influences of root distribution and growth on predicted water uptake and interspecific competition. *Oecologia* **82**, 151–157.

Fusseder, A. (1983). A method for measuring length, spatial distribution and distances of living roots *in situ*. *Plant Soil* **73**, 441–445.

Garner, S.H., Siemens, D.H. and Callaway, R.M. (1996). Tradeoffs between chemical defense and competitive ability in *Brassica rapa* (abstract). *Bull. Ecol. Soc.* **77**, 156.

Gershenzon, J. (1979). Changes in the levels of plant secondary metabolites under water and nutrient stress. *Rec. Adv. Phytochem.* **18**, 273–320.

Gianinazzi-Pearson, V. (1984). Host–fungus specificity, recognition and compatibility in mycorrhizae. In: *Genes Involved in Microbe–Plant Interactions* (Ed. by D.P.S. Verma and T. Hohn), pp. 225–253. Springer Verlag, Vienna.

Goethals, K., Mergaert, P., Geelen, D., Desomer, J., van Montagu, M. and Holsters, M. (1993). Signaling in interactions between plants and bacteria. In: *Plant Signals in Interactions with Other Organisms* (Ed. by J.C. Schultz and I. Raskin), pp. 153–163. American Society of Plant Physiologists, Rockville, MD.

Goodnight, C.J. (1985). The influence of environmental variation on group and individual selection in a cress. *Evolution* **39**, 545–558.

Gordon, D.M. and Kulig, A.W. (1996). Founding, foraging, and fighting: Colony size and the spatial distribution of harvester ant nests. *Ecology* **77**, 2393–2409.

Gordon, D.R., Welker, J.M., Menke, J.W. and Rice, K.J. (1989). Competition for soil water beween annual plants and blue oak (*Quercus douglasii*) seedlings. *Oecologia* **79**, 533–541.

Gow, N.A.R., Morris, B.M. and Reid, B. (1992). The electrophysiology of root–zoospore interactions. In: *Perspectives in Plant Cell Recognition* (Ed. by J.A. Callow and J.R. Green), pp. 173–192. Cambridge University Press, Cambridge.

Gregory, P.J. (1996). Approaches to modelling the uptake of water and nutrients in agroforestry systems. *Agrofor. Syst.* **34**, 51–65.

Grümmer, G. (1955). *Die gegenseitige Beeinflussung höherer Pflanzen—Allelopathie*. VEB Gustav Fischer Verlag, Jena.

Gulmon, S.L. and Mooney, H.A. (1977). Spatial and temporal relationships between two desert shrubs, *Atriplex hymenelytra* and *Tidestromia oblongifolia* in Death Valley, California. *J. Ecol.* **65**, 831–838.

Gur, A. and Cohen, Y. (1989). The peach replant problem—some causal agents. *Soil Biol. Biochem.* **21**, 829–834.

Gurevitch, J., Wilson, P., Stone, J.L., Teese, P. and Stoutenburgh, R.J. (1990). Competition among old-field perennials at different levels of soil fertility and available space. *J. Ecol.* **78**, 727–744.

Hall, A.B., Blum, U. and Fites, R.C. (1982). Stress modification of allelopathy of *Helianthus annuus* L. debris on seed germination. *Amer. J. Bot.* **69**, 776–783.

Hall, A.B., Blum, U. and Fites, R.C. (1983). Stress modification of allelopathy of *Helianthus annuus* L. debris on seedling biomass production of *Amaranthus retroflexus* L. *J. chem. Ecol.* **9**, 1213–1222.

Harper, J.L. (1960). Factors controlling plant numbers. In: *The Biology of Weeds* (Ed. by J.L. Harper), pp. 119–132. Blackwell Scientific Publications, Oxford.

Harper, J.L. (1961). Approaches to the study of plant competition. *Symp. Soc. exp. Biol.* **15**, 1–39.

Harper, J.L. (1985). Modules, branches, and the capture of resources. In: *Population Biology and Evolution of Clonal Organisms* (Ed. by J.B.C. Jackson, L.W. Buss and R.E. Cook), pp. 1–33. Yale University Press, New Haven.

Harper, J.L., Jones, M. and Sackville Hamilton, N.R. (1991). The evolution of roots and the problem of analysing their behaviour. In: *Plant Root Growth: An Ecological Perspective* (Ed. by D. Atkinson), pp. 3–22. Blackwell Scientific Publications, Oxford.

Harris, G.A. (1967). Some competitive relationships between *Agropyron spicatum* and *Bromus tectorum*. *Ecol. Monogr.* **37**, 89–111.

Hixon, M.A. (1980). Food production and competitor density as the determinants of feeding territory size. *Am. Nat.* **115**, 510–530.

Hook, P.B., Lauenroth, W.K. and Burke, I.C. (1994). Spatial patterns of roots in semi-arid grassland: Abundance of canopy openings and regeneration gaps. *J. Ecol.* **82**, 485–494.

Huber-Sannwald, E., Pyke, D.A. and Caldwell, M.M. (1996). Morphological plasticity following species-specific recognition and competition in two perennial grasses. *Amer. J. Bot.* **83**, 919–931.

Huber-Sannwald, E., Pyke, D.A. and Caldwell, M.M. (1997). Perception of neighbouring plants by rhizomes and roots: Morphological manifestations of a clonal plant. *Can. J. Bot.* **75**, 2146–2157.

Hutchings, M.J. and de Kroon, E.H. (1994). Foraging in plants: The role of morphological plasticity in resource acquisition. *Adv. ecol. Res.* **25**, 160–238.

Hutto, R.L., McAuliffe, J.R. and Hogan, L. (1986). Distributional associates of the saguaro (*Carnegiea gigantea*). *Southwestern Nat.* **31**, 469–476.

Inderjit, Dakshini, K.M.M. and Einhellig, F.A. (eds) (1995). *Allelopathy: Organisms, Processes, and Applications.* American Chemical Society, Washington, DC.

Israel, D.W., Giddens, J.E. and Powell, W.W. (1973). The toxicity of peach tree roots. *Plant Soil* **39**, 103–112.

Jameson, D.A. (1970). Degradation and accumulation of inhibitory substances from *Juniperus osteosperma* (Torr.) Little. *Plant Soil* **33**, 213–224.

Jones, H.G. (1993). Drought-tolerance and water-use efficiency. In: *Water Deficits: Plant Responses from Cell to Community* (Ed. by J.A.C. Smith and H. Griffiths), pp. 193–203. BIOS Scientific Publishers, Oxford.

Jones, M. and Harper, J.L. (1987a). The influence of neighbours on the growth of trees I. The demography of buds in *Betula pendula. Proc. R. Soc. Lond. B* **232**, 1–18.

Jones, M. and Harper, J.L. (1987b). The influence of neighbours on the growth of trees II. The fate of buds on long and short shoots in *Betula pendula. Proc. R. Soc. Lond. B* **232**, 19–33.

Judson, O.P. (1994). The rise of the individual-based model in ecology. *Trends Ecol. Evol.* **9**, 9–14.

Jungk, A.O. (1991). Dynamics of nutrient movement at the soil–root interface. In: *Plant Roots: The Hidden Half* (Ed. by Y. Waisel, A. Eshel and U. Kafkafi), pp. 455–481. Marcel Dekker, New York.

Kapulnik, Y., Volpin, H. and Palinski, W. (1993). Signals in plant vesicular–arbuscular mycorrhizal fungal symbiosis. In: *Plant Signals in Interactions with Other Organisms* (Ed. by J.C. Schultz and I. Raskin), pp. 142–152. American Society of Plant Physiologists, Rockville, MD.

Kawata, M. (1987). The effect of kinship on spacing among red-backed voles, *Clethrionomys rufocanus bedfordiae. Behav. Ecol. Sociobiol.* **20**, 89–97.

Kelsey, R.G. and Locken, L.J. (1987). Phytotoxic properties of cnicin, a sesquiterpene lactone from *Centaurea maculosa* (spotted knapweed). *J. chem. Ecol.* **13**, 19–33.

Koeppe, D.E., Southwick, L.M. and Bittell, J.E. (1976). The relationship between tissue chlorogenic acid concentrations and leaching of phenolics from sunflowers grown under varying phosphate nutrient conditions. *Can. J. Bot.* **54**, 593–599.

Krannitz, P.G. and Caldwell, M.M. (1995). Root growth responses of three Great Basin perennials to intra- and interspecific contact with other roots. *Flora* **190**, 161–167.

Krebs, J.R. and Davies, N.B. (1981). *An Introduction to Behavioural Ecology.* Blackwell Scientific Publishers, Oxford.

Kummerow, J., Krause, D. and Jow, W. (1977). Root systems of chaparral shrubs. *Oecologia* **29**, 163–177.

Lambin, X. and Krebs, C.J. (1993). Influences of female relatedness on the demography of Townsend's vole populations in spring. *J. Anim. Ecol.* **62**, 536–550.

Lack, D. (1954). *The Natural Regulation of Animal Numbers.* Clarendon Press, Oxford.

Lamont, B.B. and Bergl, S.M. (1991). Water relations, shoot and root architecture, and phenology of three co-occurring *Banksia* spp.: No evidence for niche differentiation in the pattern of water use. *Oikos* **60**, 291–298.

Landau, I., Muller-Scharer, H. and Ward, P.I. (1994). Influence of cnicin, a sesquiterpene lactone of *Centaurea maculosa* (Asteraceae), on specialist and generalist insect herbivores. *J. Chem. Ecol.* **20**, 929–942.

Lange, R.T. and Reynolds, T. (1981). Halo-effects in native vegetation. *Trans. R. Soc. South Austral.* **105**, 213–214.

Litav, M. and Harper, J.L. (1967). A method of studying spatial relationships between the root systems of two neighbouring plants. *Plant Soil* **26**, 389–392.

Loehwing, W.F. (1937). Root interactions of plants. *Bot. Rev.* **3**, 195–239.

Lovett, J.V. and Hoult, A.H.C. (1995). Allelopathy and self-defense in barley. In: *Allelopathy: Organisms, Processes, and Applications* (Ed. by Inderjit, K.M.M. Dakshini and F.A. Einhellig), pp. 170–183. American Chemical Society, Washington, DC.

Ludwig, J.A. (1977). Distributional adaptations of root systems in desert environments. In: *The Belowground Ecosystem: A Synthesis of Plant-associated Processes* (Ed. by J.K. Marshall), pp. 85–91. Colorado State University, Fort Collins, CO.

Lund, E.J. (1947). *Bioelectric fields and growth.* The University of Texas Press, Austin, TX.

Lynn, D.G. and Boone, L.S. (1993). Signaling germination in *Striga asiatica.* In: *Plant Signals in Interactions with Other Organisms* (Ed. by J.C. Schultz and I. Raskin), pp. 47–64. American Society of Plant Physiologists, Rockville, MD.

Lynn, D.G. and Chang, M. (1990). Phenolic signals in cohabitation: Implications for plant development. *Ann. Rev. Plant Physiol. Plant mol. Biol.* **41**, 497–526.

McKey, D., Waterman, P.G., Mbi, C.N., Gartlan, J.S. and Struhsaker, T.T. (1978). Phenolic content of vegetation in two African rain forests: Ecological implications. *Science* **202**, 61–63.

Mackie-Dawson, L.A. and Atkinson, D. (1991). Methodology for the study of roots in field experiments and the interpretation of results. In: *Plant Root Growth: An Ecological Perspective* (Ed. by D. Atkinson), pp. 25–47. Blackwell Scientific Publications, Oxford.

McAuliffe, J.R. (1988). Markovian dynamics of simple and complex desert plant communities. *Am. Nat.* **131**, 459–490.

McConnaughay, K.D.M. and Bazzaz, F.A. (1991). Is physical space a soil resource? *Ecology* **72**, 94–103.

McConnaughay, K.D.M. and Bazzaz, F.A. (1992). The occupation and fragmentation of space: Consequences of neighbouring roots. *Funct. Ecol.* **6**, 704–710.

McIntosh, R. (1992). Competition: Historical perspectives. In: *Keywords in Evolutionary Biology* (Ed. by E.F. Keller and E.A. Lloyd), pp. 61–67. Harvard University Press, Cambridge, MA.

McMichael, B.L. and Persson, H. (Eds) (1991). *Plant Roots and their Environment. Proceedings of an ISRR-Symposium, August 21–26, 1988, Uppsala, Sweden.* Elsevier, Amsterdam.

McQueen, D.R. (1968). The quantitative distribution of absorbing roots of *Pinus sylvestris* and *Fagus sylvatica* in a forest succession. *Oecol. Plant.* **3**, 83–99.

Mahall, B.E. and Callaway, R.M. (1991). Root communication among desert shrubs. *Proc. Natl Acad. Sci. USA* **88**, 874–876.

Mahall, B.E. and Callaway, R.M. (1992). Root communication mechanisms and intracommunity distributions of two Mojave Desert shrubs. *Ecology* **73**, 2145–2151.

Mahall, B.E. and Callaway, R.M. (1996). Effects of regional origin and genotype on intraspecific root communication in the desert shrub *Ambrosia dumosa* (Asteraceae). *Amer. J. Bot.* **83**, 93–98.

Maher, C.R. and Lott, D.F. (1995). Definitions of territoriality used in the study of variation in vertebrate spacing systems. *Anim. Behav.* **49**, 1581–1597.

Malmberg, T. (1980). *Human Territoriality: Survey of Behavioural Territories in Man with Preliminary Analysis and Discussion of Meaning.* Mouton Publishers, The Hague.

Mamolos, A.P., Elisseou, G.K. and Veresoglou, D.S. (1995). Depth of root activity of coexisting grassland species in relation to N and P additions, measured using non-radioactive tracers. *J. Ecol.* **83**, 643–652.

Manning, S.J. and Barbour, M.G. (1988). Root systems, spatial patterns, and competition for soil moisture between two desert shrubs. *Amer. J. Bot.* **75**, 885–893.

Mappes, T., Ylonen, H. and Viitala, J. (1995). Higher reproductive success among kin groups of bank voles (*Clethrionomys glareolus*). *Ecology* **76**, 1276–1282.

Marcum, H. and Moore, R. (1990). Influence of electrical fields and asymmetric application of mucilage on curvature of primary roots of *Zea mays*. *Amer. J. Bot.* **77**, 446–452.

Massey, A.B. (1925). Antagonism of the walnuts (*Juglans nigra* L. and *J. cinerea* L.) in certain plant associations. *Phytopathol.* **15**, 773–784.

Mead, R. (1966). A relationship between individual plant-spacing and yield. *Ann. Bot.* **30**, 301–309.

Mihaliak, C.A., Convet, D. and Lincoln, D.E. (1987). Inhibition of feeding by a generalist insect due to increased volatile leaf terpenes under nitrate-limiting conditions. *J. Chem. Ecol.* **13**, 2059–2067.

Mikola, P., Hahl, J. and Torniainen, E. (1966). Vertical distribution of mycorrhizae in pine forests with spruce undergrowth. *Ann. Bot. Fenn.* **3**, 406–409.

Milchunas, D.G. and Lauenroth, W.K. (1989). Three-dimensional distribution of plant biomass in relation to grazing and topography in the shortgrass steppe. *Oikos* **55**, 82–86.

Miller, A.L. and Gow, N.A.R. (1989a). Correlation between profile of ion-current circulation and root development. *Physiol. Plant.* **75**, 102–108.

Miller, A.L. and Gow, N.A.R. (1989b). Correlation between root-generated ion-currents, pH, fusicoccin, indoleacetic acid, and growth of the primary root of *Zea mays*. *Plant Physiol.* **89**, 1198–1206.

Mithen, R., Harper, J.L. and Weiner, J. (1984). Growth and mortality of individual plants as a function of “available area”. *Oecologia* **62**, 57–60.

Molisch, H. (1937). *Der Einfluss einer Pflanze auf die andere—Allelopathie*. Gustav Fischer, Jena.

Morris, B.M., Reid, B. and Gow, N.A.R. (1992). Electrotaxis of zoospores of *Phytophthora palmivora* at physiologically relevant field strengths. *Plant Cell Environ.* **15**, 645–653.

Morris, E.C. and Myerscough, P.J. (1991). Self-thinning and competition intensity over a gradient of nutrient availability. *J. Ecol.* **79**, 903–923.

Mou, P., Mitchell, R.J. and Jones, R.H. (1993). Ecological field theory model: A mechanistic approach to simulate plant–plant interactions in southeastern forest ecosystems. *Can. J. For. Res.* **23**, 2180–2193.

Mou, P., Jones, R.H., Mitchell, R.J. and Zutter, B. (1995). Spatial distribution of roots in Sweetgum and Loblolly Pine monocultures and relations with above-ground biomass and soil nutrients. *Funct. Ecol.* **9**, 689–699.

Muller, C.H. (1946). *Root Development and Ecological Relations of Guayule*. United States Department of Agriculture, Washington, DC.

Muller, C.H. (1969). Allelopathy as a factor in ecological process. *Vegetatio* **18**, 348–357.

Muller, C.H., Muller, W.H. and Haines, B.L. (1964). Volatile growth inhibitors produced by aromatic shrubs. *Science* **143**, 471–473.

Nakamura, R.R. (1980). Plant kin selection. *Ecol. Theor.* **5**, 113–117.

Nicholson, A.J. (1954). An outline of the dynamics of animal populations. *Austral. J. Zool.* **2**, 9–65.

Nilsson, M.-C. (1994). Separation of allelopathy and resource competition by the boreal dwarf shrub *Empetrum hermaphroditum* Hagerup. *Oecologia* **98**, 1–7.

Nobel, P.S. (1981). Spacing and transpiration of various sized clumps of a desert grass, *Hilaria rigida*. *J. Ecol.* **69**, 735–742.
Nobel, P.S. (1997). Root distribution and seasonal production in the northwestern Sonoran Desert for a C_3 subshrub, a C_4 bunchgrass, and a CAM leaf succulent. *Amer. J. Bot.* **84**, 949–955.
Nobel, P.S. and Franco, A.C. (1986). Annual root-growth and intraspecific competition for a desert bunchgrass. *J. Ecol.* **74**, 1119–1126.
Northup, R.R., Yu, Z., Dahlgren, R.A. and Vogt, K.A. (1995). Polyphenol control of nitrogen release from pine litter. *Nature* **377**, 227–229.
Nye, P.H. and Tinker, P.B. (1977). *Solute Movement in the Soil–Root System*. Blackwell Scientific Publications, Oxford.
Oleszek, W. (1987). Allelopathic effects of volatiles from some Cruciferae species on lettuce, barnyard grass and wheat growth. *Plant Soil* **102**, 271–273.
Parrish, J.A.D. and Bazzaz, F.A. (1976). Underground niche separation in successional plants. *Ecology* **57**, 1281–1288.
Persson, H. (1980). Spatial distribution of fine-root growth, mortality and decomposition in a young Scots pine stand in Central Sweden. *Oikos* **34**, 77–87.
Ponder, F., Jr and Tadros, S.H. (1985). Juglone concentration in soil beneath black walnut interplanted with nitrogen-fixing species. *J. Chem. Ecol.* **11**, 937–942.
Provencher, L. and Vickery, W. (1988). Territoriality, vegetation complexity, and biological control: The case for spiders. *Am. Nat.* **132**, 257–266.
Putnam, A.R. and Weston, L.A. (1986). Adverse impacts of allelopathy in agricultural systems. In: *The Science of Allelopathy* (Ed. by A.R. Putnam and C.-S. Tang), pp. 43–56. John Wiley & Sons, New York.
Raper, C.D., Jr and Barber, S.A. (1970). Rooting systems of soybeans. I. Differences in root morphology among varieties. *Agron. J.* **62**, 581–584.
Reinartz, J.A. and Popp, J.W. (1987). Structure of clones of northern prickly ash (*Xanthoxylum americanum*). *Amer. J. Bot.* **74**, 415–428.
Rice, E.L. (1984). *Allelopathy*. Academic Press, Orlando.
Rice, E.L. (1987). Allelopathy: An overview. In: *Allelochemicals: Role in Agriculture and Forestry* (Ed. by G.R. Waller), pp. 8–22. American Chemical Society, Washington, DC.
Rice, E.L. (1995). *Biological Control of Weeds and Plant Diseases: Advances in Applied Allelopathy*. University of Oklahoma Press, Norman.
Richards, D. and Rowe, R.N. (1977). Effects of root restriction, root pruning and 6-benzylaminopurine on the growth of peach seedlings. *Ann. Bot.* **41**, 729–740.
Robberecht, R., Mahall, B.E. and Nobel, P.S. (1983). Experimental removal of intraspecific competitors—effects on water relations and productivity of a desert bunchgrass, *Hilaria rigida*. *Oecologia* **60**, 21–24.
Rogers, W.S. and Head, G.C. (1969). Factors affecting the distribution and growth of roots of perennial woody species. In: *Root Growth* (Ed. by W.J. Whittington), pp. 280–292. Butterworths, London.
Rosenthal, G.R. and Janzen, D.H. (eds) (1979). *Herbivores: Their Interaction with Plant Secondary Metabolites*. Academic Press, New York.
Rovira, A.D. (1969). Plant root exudates. *Bot. Rev.* **35**, 35–57.
Sackville Hamilton, N.R., Jones, M. and Harper, J.L. (1991). Automated analysis of roots in bulk soil. In: *Plant Root Growth: An Ecological Perspective* (Ed. by D. Atkinson), pp. 69–73. Blackwell Scientific Publishers, Oxford.
St John, T.V. (1983). Response of tree roots to decomposing organic matter in two lowland Amazonian rain forests. *Can. J. For. Res.* **13**, 346–349.

St John, T.V., Coleman, D.C. and Reid, C.P.P. (1983). Growth and spatial distribution of nutrient-absorbing organs: Selective exploitation of soil heterogeneity. *Plant Soil* **71**, 487–493.

Schenk, H.J. (1999). Clonal splitting in desert shrubs. *Plant Ecol.* (in press).

Schrank, A.R. (1959). Electronasty and electrotropism. In: *Movements due to Mechanical and Electrical Stimuli and to Radiation* (Ed. by W. Ruhland), pp. 148–163. Springer Verlag, Berlin.

Schreiner, O. and Reed, H.S. (1907). The production of deleterious excretions by roots. *Bull. Torr. bot. Club* **34**, 279–303.

Schmitt, T.M., Hay, M.E. and Lindquist, N. (1995) Constraints on chemically mediated coevolution: Multiple functions for seaweed secondary metabolites. *Ecology* **76**, 107–123.

Sheik, K.H. and Rutter, A.J. (1969). The response of *Molinia caerulea* and *Erica tetralix* to soil aeration and related factors. I. Root distribution in relation to soil porosity. *J. Ecol.* **57**, 713–726.

Silander, J.A., Jr and Pacala, S.W. (1990). The application of plant population dynamics models to understanding plant competition. In: *Perspectives on Plant Competition* (Ed. by J.B. Grace and D. Tilman), pp. 67–92. Academic Press, San Diego.

Singh, S.P. (1964). *Cover, Biomass, and Root-shoot Habit of* Larrea divaricata *on a Selected Site in Southern New Mexico*. Thesis New Mexico State University, Las Cruces.

Smith, M.T. and Van Staden, J. (1995). Infochemicals: The seed–fungus–root continuum. A review. *Environ. exp. Bot.* **35**, 113–123.

Soriano, A., Golluscio, R.A. and Satorre, E. (1987). Spatial heterogeneity of the root system of grasses in the Patagonian arid steppe. *Bull. Torr. bot. Club* **114**, 103–108.

Southon, T.E. and Jones, R.A. (1992). NMR imaging of roots: Methods of reducing the soil signal and for obtaining a 3-dimensional description of the roots. *Physiol. Plant.* **86**, 322–328.

Sprugel, D.G., Hinckley, T.M. and Schaap, W. (1991). The theory and practice of branch autonomy. *Ann. Rev. Ecol. Syst.* **22**, 309–334.

Stenz, H.-G. and Weisenseel, M.H. (1991). DC-electric fields affect the growth direction and statocyte polarity of root tips (*Lepidium sativum*). *J. Plant Physiol.* **138**, 335–344.

Stenz, H.-G. and Weisenseel, M.H. (1993). Electrotropism of maize (*Zea mays* L.) roots: Facts and artifacts. *Plant Physiol.* **101**, 1107–1111.

Stevens, K.L. (1986). Allelopathic polyacetylenes from *Centaurea repens* (Russian knapweed). *J. Chem. Ecol.***12**, 1205–1211.

Stevens, L., Goodnight, C.J. and Kalisz, S. (1995). Multilevel selection in natural populations of *Impatiens capensis*. *Am. Nat.* **145**, 513–526.

Stocker, O. (1928). Der Wasserhaushalt ägyptischer Wüsten- und Salzpflanzen vom Standpunkt einer experimentellen und vergleichenden Pflanzengeographie aus. *Bot. Abhdlg.* **13**, 200 pp.

Story, R. (1967). Pasture patterns and associated soil water in partially cleared woodland. *Austral. J. Bot.* **15**, 175–187.

Tang, C.-S., Cai, W.-F., Kohl, K. and Nishimoto, R.K. (1995). Plant stress and allelopathy. In: *Allelopathy: Organisms, Processes, and Applications* (Ed. by Inderjit, K.M.M. Dakshini and F.A. Einhellig), pp. 142–157. American Chemical Society, Washington, DC.

Thomas, R.B. and Strain, B.R. (1991). Root restriction as a factor in photosynthetic acclimation of cotton seedlings grown in elevated carbon dioxide. *Plant Physiol.* **96**, 627–634.

Tilman, D. (1989). Competition, nutrient reduction, and the competitive neighborhood of a bunchgrass. *Funct. Ecol.* **3**, 215–219.

Tonsor, S.J. (1989). Relatedness and intraspecific competition in *Plantago lanceolata*. *Am. Nat.* **134**, 897–906.

Tschaplinski, T.J. and Blake, T.J. (1985). Effects of root restriction on growth correlations, water relations and senescence of alder seedlings. *Physiol. Plant.* **64**, 167–176.

Upton, G.J.G. and Fingleton, B. (1985). *Spatial Data Analysis by Example. Vol. 1: Point Patterns and Quantitative Data*. John Wiley & Sons, Chichester.

Vogt, K.A., Vogt, D.J., Moore, E.E. and Sprugel, D.G. (1989). Methodological considerations in measuring biomass, production, respiration and nutrient resorption for tree roots in natural ecosystems. In: *Applications of Continuous and Steady-state Methods to Root Biology* (Ed. by J.G. Torrey and L.J. Winship), pp. 217–232. Kluwer Academic Publishers, Dordrecht.

Weaver, J.E. (1919). *The Ecological Relations of Roots*. Carnegie Institution of Washington, Washington, DC.

Weaver, J.E. (1920). *Root Development in the Grassland Formation*. Carnegie Institution of Washington, Washington, DC.

Webb, L.J., Tracey, J.G. and Haydock, K.P. (1967). A factor toxic to seedlings of the same species associated with living roots of the non-gregarious subtropical rain forest tree *Grevillea robusta*. *J. appl. Ecol.* **4**, 13–25.

Welden, C.W., Slauson, W.L. and Ward, R.T. (1990). Spatial pattern and interference in piñon-juniper woodlands of northwest Colorado. *Great Basin Nat.* **50**, 313–319.

Went, F.W. (1955). The ecology of desert plants. *Sci. Am.* **192**, 68–75.

Went, F.W. (1970). Plants and the chemical environment. In: *Chemical Ecology* (Ed. by E. Sondheimer and J.B. Simeone), pp. 71–82. Academic Press, New York.

Whittaker, R.H. (1970). The biochemical ecology of higher plants. In: *Chemical Ecology* (Ed. by E. Sondheimer and J.B. Simeone), pp. 43–70. Academic Press, New York.

Whittaker, R.H. and Feeny, P.P. (1971). Allelochemics: Chemical interactions between species. *Science* **171**, 757–770.

Williams, G.C. (1966). *Adaptation and Natural Selection: A Critique of Some Current Evolutionary Thought*. Princeton University Press, Princeton, NJ.

Williamson, G.B. (1990). Allelopathy, Koch's postulates, and the neck riddle. In: *Perspectives on Plant Competition* (Ed. by J.B. Grace and D. Tilman), pp. 143–162. Academic Press, San Diego.

Willson, M.F., Thomas, P.A., Hoppes, W.G., Katusic-Malmborg, P.L., Goldman, D.A. and Bothwell, J.L. (1987). Sibling competition in plants: An experimental study. *Am. Nat.* **129**, 304–311.

Wilson, D.S. (1980). *The Natural Selection of Populations and Communities*. The Benjamin/Cummings Publishing Company, Inc., Menlo Park, CA.

Wilson, E.O. (1975). *Sociobiology: The New Synthesis*. The Belknap Press of Harvard University Press, Cambridge, MA.

Wilson, J.B. (1987). Group selection in plant populations. *Theor. appl. Genet.* **74**, 493–502.

Wilson, J.B. and Newman, E.I. (1987). Competition between upland grasses: Root and shoot competition between *Deschampsia flexuosa* and *Festuca ovina*. *Acta Oecol. Oecol. Gen.* **8**, 501–509.

Woods, F.W. (1960). Biological antagonisms due to phytotoxic root exudates. *Bot. Rev.* **26**, 546–569.

Wright, R.A. and Mueller-Dombois, D. (1988). Relationships among shrub population structure, species association, seedling root form and early volcanic succession. In: *Plant Form and Vegetation Structure* (Ed. by M.J.A. Werger, P.J.M. van der Aart, H.J. During and J.T.A. Verhoeven), pp. 87–104. SPB Academic Publishing, The Hague.
Wynne-Edwards, V.C. (1962). *Animal Dispersion in Relation to Social Behaviour.* Oliver and Boyd, Edinburgh.
Wynne-Edwards, V.C. (1986). *Evolution Through Group Selection.* Blackwell Scientific Publishers, Oxford.

The Relationship between Animal Abundance and Body Size: A Review of the Mechanisms

T.M. BLACKBURN AND K.J. GASTON

I. SUMMARY

An interspecific relationship between animal abundance and body size is one of the most frequently reported patterns in ecology. However, there has been little progress towards an understanding of the mechanisms causing the relationship, focus having dwelt on debate over its form. Only one mechanism has been given serious consideration as an explanation for abundance–body size relationships. This is that abundances are constrained by energy availability. However, there are at least five other hypotheses in the literature. Here, we critically assess the extent to which each hypothesis can explain the observed patterns. None of the six hypotheses can be considered adequate as explanations of the abundance–body size relationship. Most importantly, the energetic constraint hypothesis is shown to be logically flawed because, if its predictions hold, then its assumptions must be incorrect (and vice versa). Moreover, it is never possible to prove that abundances are energy-limited. We suggest a new, and more general, model whereby a

ADVANCES IN ECOLOGICAL RESEARCH VOL. 28
ISBN 0–12–013928–6

negative abundance–body size relationship arises as a logical consequence of the observed frequency distribution of biomasses amongst species, and a constraint on minimum viable abundance. However, explanations for what shapes the biomass distribution and how minimum viable abundances are related to body size are both likely to be difficult to obtain.

II. INTRODUCTION

The relationship between abundance and body size has been one of the most extensively studied interspecific patterns in animal ecology (see reviews in Cotgreave, 1993; Blackburn and Gaston, 1997a; Blackburn and Lawton, 1994). It has been documented in most major environments and taxa, and across spatial scales ranging from single trees and ponds to entire continents (Damuth, 1981, 1987; Gaston, 1988; Morse *et al.,* 1988; Basset and Kitching, 1991; Blackburn *et al.,* 1993a; Solonen, 1994; Strayer, 1994; Cotgreave, 1995; Gregory and Blackburn, 1995; Silva and Downing, 1995; Gaston and Blackburn, 1996a; Greenwood *et al.,* 1996). Surprisingly, however, the large amount of attention that the relationship has received has not resulted in consensus as to its precise form (for a recent review, see Blackburn and Gaston, 1997a). Debate has focused on whether the abundance–body size relationship has a simple linear negative form or is more complicated. It has also persisted at the cost of critical consideration of the mechanisms that are likely to generate such relationships. Indeed, by and large, the treatment of evidence for possible mechanisms has progressed little beyond simple statements of the particular form of the relationship in any given study.

The primary reason for this lack of progress is that the single most influential study of the interspecific abundance–body size relationship (Damuth, 1981) presented not only a strong pattern, but also a seductive mechanism to accompany it. Damuth compiled data from the literature on body mass and population density for 307 species of mammalian terrestrial primary consumers (Figure 1). He found that the regression of population density (D) on body mass (M), with both variables logarithmically transformed, gave a negative relationship of slope – 0.75 across these species, and that body mass explained a high proportion of the variance in density ($r^2 = 0.74$). This corresponds to the power relationship $D = cM^{-0.75}$, where c is a constant. The slope of the relationship indicated a potential underlying mechanism. Metabolic rate (R) increases with body mass in mammals according to the power relationship $R \propto M^{0.75}$ (Kleiber, 1962; Peters, 1983; Elgar and Harvey, 1987; Nagy, 1987; Reiss, 1989). Therefore, the total energy used by a local population of mammalian primary consumers per unit time (the product of the population density and the metabolic requirements of each individual) appears to be independent of body size: that is, $D \bullet R \propto M^{-0.75} \bullet M^{0.75} \propto M^{0.0}$. This result leads to the somewhat unexpected

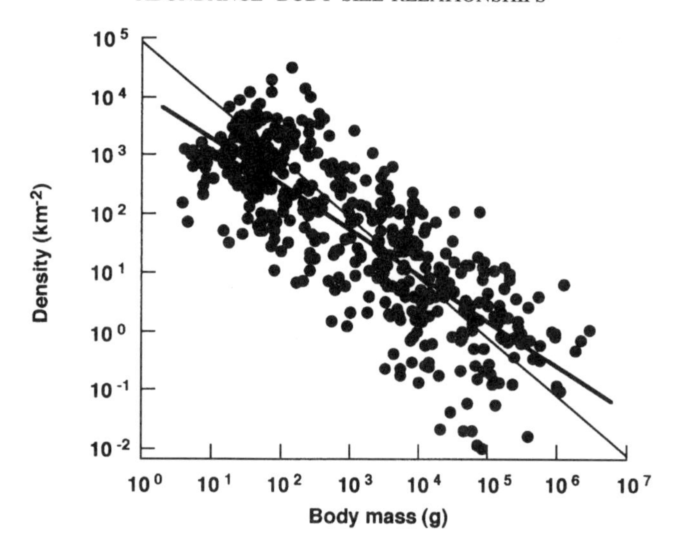

Fig. 1. The relationship between density (numbers per km²) and body mass (g) for a selection of mammals of the world (from data in Damuth, 1987) ($n = 467$). The position of the OLS regression slope is indicated by the thick line, which has slope – 0.75. For reference, the thin line has a slope of –1.

conclusion that species of different body sizes in an assemblage use approximately equal amounts of energy (Damuth, 1981). This has since been dubbed the "energetic equivalence rule" (Nee *et al.,* 1991b).

Subsequent to Damuth's study, any negative abundance–body mass relationship with a slope not differing significantly from – 0.75 has generally been taken as evidence in support of the energetic equivalence rule. Any other relationship has been taken as evidence against it. Critical assessments of other assumptions and predictions of the hypothesis have by and large been ignored (but see Lawton, 1989, 1990; Blackburn and Lawton, 1994; Marquet *et al.,* 1995). Further, the focus on this single mechanism has hampered the exploration of other possibilities. Yet, we are aware of six explanations for abundance–body size relationships that exist, at least implicitly, in the literature. While these hypotheses are not generally incompatible, some have serious implications for the evidence usually regarded as providing support for others. In this paper, we critically review the different mechanisms that might explain the interspecific relationship between abundance and body size in animals.

Four of the six hypotheses we address here have been formulated to explain abundance–body size relationships across species. The remaining two have been suggested to explain patterns observed when accounting for phylogenetic relatedness amongst species (hereafter, "within taxa"). Ultimately, the strongest evidence for hypotheses about what causes ecological relationships will come from tests that consider the evolutionary association between variables and hence from tests that take account of phylogenetic relatedness. This is because ecologists are interested in traits that affect species' responses to current ecological circumstances, but the distribution of these traits across species depends on evolutionary relationships (Harvey, 1996). However, initially at least, we assess how well each hypothesis explains those patterns for which it was proposed as a mechanism. Specifically, we cover mechanisms that deal with effects of census area, energy constraints, latitudinal gradients, concatenation from abundance and body size distributions, interspecific competition and differential extinction. Each in turn is first defined and then critically evaluated with respect to their utility for explaining observed empirical relationships between abundance and body size. Finally, given that none of the mechanisms we review is entirely adequate as an explanation of abundance–body size relationships, we discuss features of such relationships that may benefit from further consideration. For the most part, we assume that the abundance–body size relationship is, to some degree, negative. The issue of the form of the relationship is discussed at length by Blackburn and Gaston (1997a); at large spatial scales, at least, a negative relationship is a reasonable assumption.

III. MECHANISMS

The hypothesised mechanism most commonly considered to generate the relationship between abundance and body size is that the abundance of a species is a consequence of the energetic requirements of its individuals. Before we address this proposition in detail, however, it is necessary to examine one of the other possible causes of the pattern—that the relationship is an artefact of the way abundances are measured. This mechanism must be considered first because it has implications that potentially affect the assessment of all other hypotheses.

A. The Census Area Mechanism

1. Description

The census area mechanism only applies to the abundance–body size relationship when abundance is expressed in terms of density. Density is normally measured as the number of individuals (or some other relevant unit) in a given area. Thus it has two components, the number of animals and the census area. Variation in either component can affect the observed density of a species. The census area mechanism suggests that a broadly negative interspecific relationship between density and body size arises because small-bodied species are censused across smaller areas.

While much attention has been focused on methods of determining accurately what is the actual number of individuals for a density estimate (e.g. Seber,

1982), almost none has been given to the question of the efffect of the area over which the density estimate is obtained (but see Haila, 1988; Gaston, 1994). Consider the interspecific relationship between body mass and abundance described in Figure 1, plotted from the data given in Damuth (1987). This plot includes densities of mice and elephants. These densities are unlikely to have been obtained from censuses over areas of similar size. In fact, within these data there is a strong positive relationship between the body mass of a species and the area over which its density was censused (Blackburn and Gaston, 1996a, 1997b). A similar relationship has been shown for mammalian carnivores (Schonewald-Cox *et al.,* 1991; Smallwood and Schonewald, 1996). For the mammalian primary consumers in Figure 1, the area over which a species is censused is a better (in terms of coefficient of determination) predictor of its abundance than is its body mass (Blackburn and Gaston, 1996a, 1997b).

The interspecific relationship between density and body size could be the result of the variance in census area. This explanation additionally requires only that densities tend to be measured in areas where a species occurs. An example of how it could work is given in Figure 2. The two species in this hypothetical area both occur at the same density, but their abundances are estimated using census areas of different size. With the census areas positioned as in the figure, the large-bodied species has an estimated density of 12 km^{-2} and the small-bodied species an estimated density of 32 km^{-2}. Given the positioning of the individuals in Figure 2, the maximum density that it is possible to record is 48 km^{-2} for the small-bodied species and 16 km^{-2} for the large-bodied species. The absolute range of densities for the small-bodied species would occur if the census area either contained all the individuals present in the 1 km^2, or only one, giving a range of 16–160 km^{-2}. The same range for the large-bodied species is 4–40 km^{-2}. In this simple example, a lower density is likely to be recorded for the large-bodied species (and note that the true density of the small-bodied species will only ever be approached if areas with no individuals are included in the estimate!).

Many factors determine why an area of a particular size is chosen for study of a given species. Smaller areas will be favoured for practical reasons of relative ease of sampling, delineation of study area, control of disturbance and replication. Conversely, areas must in general be large enough that sufficient numbers of individuals occur within them (species are seldom studied in areas in which they are difficult to find), and that populations are not dominated by transient individuals. Trade-offs between these factors are likely to result in different-sized census areas for different-sized species (as well as species with different kinds of population dynamics, trophic habits, habitat usage, etc.). The allometry of density could arise from circular logic, whereby body size influences the spatial extent of study, and the resulting density estimates are related back to body size (Smallwood *et al.,* 1996).

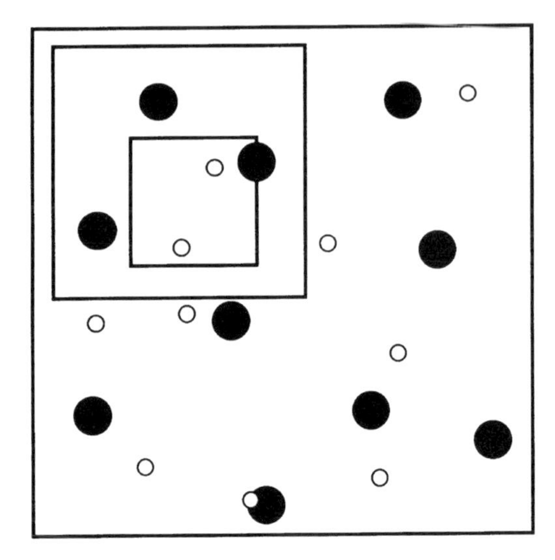

Fig. 2. A hypothetical area of land. Assume that the large square has an area of 1 km^{-2}. One large-bodied and one small-bodied species occur in this area, with territories represented by the large filled and small unfilled circles, respectively. Both occur at the same density, namely 10 km^{-2}. However, the abundance of the large-bodied species is censused across an area of 0.25 km^2 (the medium-sized square), while that of the small-bodied species is represented censused across an area of 0.0625 km^2 (the smallest square illustrated).

2. *Evaluation*

The census area mechanism alone is probably insufficient to account for the full range of variation observed in the abundances of organisms. Few would argue that the densities observed for a set of species are entirely the consequence of the area over which they are sampled. At the extreme, nematodes in the soil really do occur at higher densities than the elephants treading it. Census areas are not chosen without reference to the likely density of the species of interest. However, there is clear evidence that density and census area are not independent. Over moderate to large areas, the densities of individual species should tend to decline as census area increases, because more sub-areas will be included in which individuals do not occur. Hence, there is no such thing as a single density for a species. Rather, each species has its own (intraspecific) relationship between density and census area (Blackburn and Gaston, 1996a, 1997b). The slopes of these relationships are likely to, and indeed do, differ between species, but whenever they differ from zero, density and census area will not be independent.

Given that density and census area are not independent, the question arises of whether census areas for species of different size are comparable, in the sense that they result in measurement of equivalent densities (an issue with far wider ramifications than simply for abundance–body size relationships). There are reasons to doubt that they are comparable in this sense. First, it would be necessary that the trade-offs between the various factors determining the size of a study area result in ecologically equivalent areas being used for the measurement of density both for small and large species. As these factors are predominantly determined by methodological, rather than biological, considerations, this seems exceedingly unlikely.

Secondly, density estimates generally tend to be derived from increasingly site-oriented rather than species-oriented studies as the body size of the species studied increases. A high proportion of studies which include density estimates for large mammal species, for example, are studies of the large mammal faunas of particular parks or preserves. Densities derived from site-oriented studies typically pay less attention to the patterns of usage of the site by different species than do species-oriented studies. For example, densities may be calculated for a park or reserve for species of a range of body sizes (e.g. dik-dik to elephant), on the basis of precisely the same value of area (e.g. the size of the park). Densities of species that use restricted parts of this area hence will be underestimated (perhaps drastically) relative to species using a greater proportion. This suggests the possibility that there may be systematic differences in the relative use made of census areas by animals of different body sizes. Replotting the abundance–body size relationship for mammalian primary consumers, statistically controlling for the average area over which density was censused for each species, yields a relationship that is still negative, but with both the slope and correlation coefficient greatly reduced (Blackburn and Gaston, 1996a). The clear implications are that different kinds of densities are being measured for large and small animals, and that the interpretation of patterns of abundance from such interspecific comparisons will be confounded by uncertainty as to what is actually being compared. It is by no means clear how to obtain equivalent densities for interspecific comparisons (Blackburn and Gaston, 1996a, 1997b).

In summary, the census area mechanism alone is unlikely to generate the negative abundance–body size relationship, at least across very large ranges of body sizes. Nevertheless, it does strongly suggest that the negative relationships typically recovered from studies where abundance data are compiled from a wide range of studies using a wide range of census methodologies (e.g. Figure 1) may in part be artefacts of the data used (Blackburn and Gaston, 1997b). This needs to be borne in mind when considering subsequent mechanisms. It is to these mechanisms that we now turn our attention.

B. The Energetic Constraint Mechanism

1. Description

Metabolic rate (R) increases with body mass (M) in animals approximately according to the power function $R = cM^y$, where c is a taxon-specific constant.

The exponent y is typically assumed to be 0.75, although a range of values have been reported across taxa (Table 1; Kleiber, 1962; Peters, 1983; Elgar and Harvey, 1987; Nagy, 1987; Reiss, 1989). This interspecific scaling relationship indicates that large-bodied animals have smaller mass-specific metabolic requirements, but larger total metabolic demands, than do small-bodied animals. Thus it takes absolutely more energy to maintain an individual of a large-bodied species, for a given unit of time, than it does to maintain an individual of a small-bodied species. However, the amount of energy available in the environment at a given point in space and time is finite; therefore, so is the number of individual animals that the environment can support. It follows that species comprising large-bodied individuals must either manage to appropriate a higher proportion of the overall amount of energy available or have fewer individuals per unit area than species comprising small-bodied individuals. If the former is true, then the abundance–body size relationship is not logically restricted to any particular form; if the latter is true, however, it is constrained to be negative.

Extending the preceding reasoning, the slope of the abundance–body size relationship should give information on the extent to which species of different sizes appropriate resources. The total energy used by a local population of animals per unit time is the product of the population's density (D) and the metabolic requirements (R) of each individual. Assuming $R \propto M^{0.75}$, how population energy use ($D \bullet R$) scales with body mass depends on the exponent x of the relationship between mass and density, $D \propto M^{x}$, so that $D \bullet R \propto M^{0.75}M^{x} \propto M^{y}$. Simply, if $x = -0.75$, then $y = 0$, and population energy use is independent of body mass. If $x > -0.75$ then y is positive, and populations of large-bodied species tend to utilise a higher proportion of the available energy than do small-bodied species (vice versa if $x < -0.75$).

2. Evaluation

As discussed in section II, the first empirical examination of the allometric scaling of population density that specifically considered the distribution of

Table 1
Minimum and maximum values of y given by Peters (1983; Appendix IIIa and IIIB) for the equation $R = cM^{y}$ for different taxa, where R = metabolic rate, M = body mass, and c is a taxon specific constant

Taxon	Minimum	Maximum
Mammals	0.610	0.790[1]
Birds	0.476	0.865
Reptiles	0.600	0.880
Amphibians	0.622	0.860
Insects	0.620	0.860
Crustacea	0.435	0.850

[1] 1.02 for hibernating mammals.

energy use found that $x = -0.75$ across species of mammalian primary consumers (Damuth, 1981). Thus, mammals of different body sizes appear to use approximately equal amounts of energy. Therefore, assuming that equal amounts of energy are available to species of different body sizes, abundances seem to be constrained by energy availability (Damuth, 1981). Blackburn and Gaston (1997a) showed that ordinary least squares (OLS) regression slopes for abundance–body mass relationships tend to cluster around -0.75, although the scatter is wide and the mean slope is -0.51. Broadly, the energetic equivalence rule (EER) seems to hold.

The idea that all animals appropriate approximately equal amounts of energy from the environment, and thus that the density an animal attains is constrained by energy availability to be a simple inverse function of its body mass, is appealing. However, it essentially ignores a number of critical issues.

(1) The allometric scaling of population energy use is obtained, almost without exception, by summing the allometric exponents of metabolic rate and population density (e.g. Damuth, 1981; Brown and Maurer, 1986; Silva and Downing, 1995). The calculation of $D{\bullet}R$ above is one such example. The resulting exponent (y where $D{\bullet}R \propto M^y$) is then the product of the mean component values, rather than the mean of their products (Welsh *et al.*, 1988; Medel *et al.*, 1995). Combining exponents assumes that these quantities will be the same; in other words, it assumes that y is the exponent that would be obtained if body mass was regressed against population energy use calculated separately for each species. This assumption may be incorrect because it ignores co-variation between the component variables. The exponent y will only be an unbiased estimate of the allometric scaling of population energy use if metabolic rate and density are themselves uncorrelated. Since metabolic rate is argued in part to determine density, this seems unlikely. This assumption is a specific example of a general phenomenon known as the "fallacy of averages" (Welsh *et al.*, 1988, after Wagner, 1969). A procedure to correct for this potential problem is given by Welsh *et al.* (1988).

(2) Focusing on the value of the regression slope ignores the variance around it (Blackburn and Lawton, 1994). Since the slope is derived from logarithmically transformed data, this variance may be high even for quite strong linear negative relationships; there are close to four orders of magnitude of variation in density for mammals of a given mass (Figure 1). Two species of equal mass differing in density by three orders of magnitude would be predicted, under the EER, to have similarly different metabolic rates. If they had the same metabolic rate, then their population energy use would also show three orders of magnitude difference, which would not be good evidence for an EER. A simple test of the EER would be to see whether residual variation in the abundance–body mass relationship for any taxon can be explained by metabolic rate differences. The single analysis to date suggests that it cannot (Blackburn *et al.*, 1996).

Kozlowski and Weiner (1996) note that there are two ways in which allometric relationships, such as that between abundance and body size, can be explained. First, they may reflect functional relationships between body size and the variable of interest. Secondly, they may represent a by-product of some underlying mechanism. In the first case, residual variation is treated as error caused by other random effects. However, as Kozlowski and Weiner point out, the large magnitude of residual error around many allometric relationships belies this assumption, and calls into question any direct functional interpretation. Since the energetic equivalence rule is arguably a functional explanation for the allometry of density, these arguments suggest that it is inadequate. The lack of a density–body size relationship within taxa (see below) also suggests that the functional interpretation cannot apply.

(3) Even if a slope of – 0.75 is found to be representative of the relationship (see above), other mechanisms than the EER may be responsible. For example, Marquet *et al.* (1990) obtained slopes close to – 0.75 for rocky shore invertebrate assemblages where space, not energy, was the factor limiting abundance.

Further, a slope of – 0.75 for the abundance–body mass relationship does not prove that energetic constraints determine abundances. Rather, it shows that all species use about the same amount of energy. While at first sight this seems to amount to the same thing, it does not. Consider the following example. Population energy use scales with body mass to the 0.25 power in a hypothetical animal assemblage; large-bodied species use a greater total amount of energy than do small-bodied species. However, the total amount of energy available to a species also scales with body mass to the 0.25 power. The densities of species are constrained by the amount of energy available to them, *but that cannot be concluded from the slope of the relationship between population energy use and body size.* The conclusion that the EER indicates that abundances are energy limited arises solely from the assumption that equal amounts of energy are available to species of all body masses. This assumption is critical, seems unlikely to be true (Brown and Maurer, 1986, 1989; Harvey and Lawton, 1986; Lawton, 1990), but remains untested. Until a test is performed, the foundations of the EER must be considered weak.

One good reason to believe that equal amounts of energy are not available to species of different body sizes is that the frequency distribution of body masses is not uniform (reviews in May, 1978; Blackburn and Gaston, 1994; Brown, 1995). Assume that the EER is true; therefore, the average amount of energy used by species is independent of their body mass. However, if the mean is independent of mass, the total amount of energy used by species will not be. Rather, it will be proportional to the number of species of a given mass. If the EER is true, the total amount of energy available to species of a given body mass must be proportional to the number of species of that size, and therefore energy is not equally available to species of all sizes. If energy is not equally available to species of all sizes, an abundance–body mass relationship of slope – 0.75 does not imply an energetic constraint (Figure 3).

This suggests that the EER does not provide evidence that abundances are energetically constrained. It only indicates that abundances are energy limited if equal amounts of energy are available to species of all body masses. Yet if the EER holds, then equal amounts of energy are not available to species of all body masses! (Note that this argument only does not apply in the unlikely case of a completely uniform body mass distribution.) Energetic limitation of abundances therefore can only apply if equal amounts of energy are available to all species individually (i.e. regardless of how many other species of similar body size there are), but if this was the case, we would not expect the observed wide variation in abundances within a group of species of similar body size. If energetic limitation does apply, it should not produce a slope of

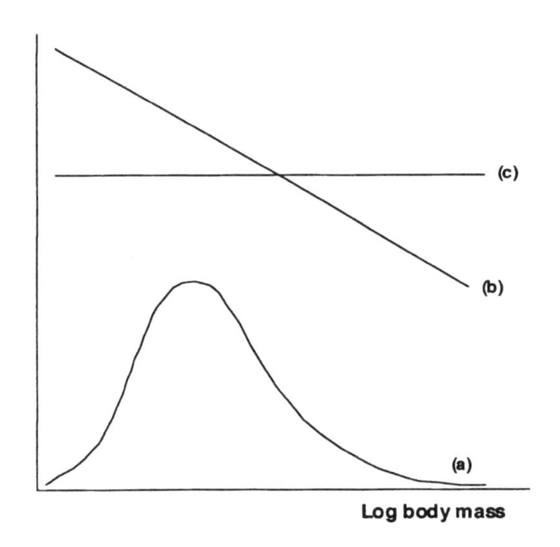

Fig. 3. A graphical illustration of a logical flaw in the hypothesis that abundances are constrained by the energy available to species of different body sizes. The lines represent: (a) the frequency distribution of log body masses; (b) the relationship between log abundance and log body mass; and (c) the amount of energy available to species of different body mass. Assuming (c) to be true, then a slope of – 0.75 for (b) has been taken as indicating energetic equivalence (assuming that metabolic rate scales as $M^{0.75}$). However, if energetic equivalence is correct, then the amount of energy used by each species is equal, and the total amount of energy used by species of a given body mass should parallel (a). Then, either the amount of energy used by species of a given body mass is equivalent to the amount available, in which case (c) cannot be true (it should resemble (a)), or the amount of energy used is not equivalent to the amount available (both (a) and (c) are correct), in which case (c) cannot constrain abundances.

– 0.75 in the abundance–body mass relationship, because some sizes have more energy available to them than others.

If energy does not limit the abundances of species in general, it has been suggested that it could still limit the abundances of the most common species (Lawton, 1990; Blackburn *et al.,* 1993a,b); that is, those species near the upper boundary of negative portions of abundance–body size plots. Blackburn *et al.* (1993a,b) argued that the evidence counted against this suggestion, because the upper bound slopes of plots from both sample and compilation studies were generally more steeply negative than the – 0.75 predicted by energetic arguments. However, as we have shown above, energetic constraint arguments only predict slopes of – 0.75 if equal amounts of energy are available to species of all body masses. If more energy were available to small-bodied species, then steeper upper bound slopes would be expected. Further, if the amount of energy available to a taxon was reflected in the amount appropriated by that taxon, then the upper bound slope expected would depend on the shape of the frequency distribution of body mass and the slope of the abundance–body mass pattern. In the case of Damuth's mammal data (Figure 1), the slope of – 0.75 implies equal average energy use, but there are more small than large-bodied species. The upper bound slope should be steeper than – 0.75, and in fact it is – 0.96 (Blackburn *et al.,* 1993a).

However, shifting energetic constraint arguments from abundances in general to maximum abundances does not circumvent any of the problems associated with energy availability to different-sized animals. In fact, it only complicates matters. In particular, there is no logical reason why the slope of maximum abundances should reflect the amount of energy available to species of different body sizes. To do so, the most abundant species of each body size would have to appropriate the same proportion of the energy available to all species of that body size. Once again, the fact that there are more small-bodied than large-bodied species makes this extremely unlikely; under the EER, where there are more species, each would be expected to get a smaller proportion of available resources.

Another complication is that the above arguments assume that the amount of energy available to species of different sizes is reflected in the amounts of energy that those species actually appropriate. Yet, again, there is no logical necessity for this to be true. Equal amounts of energy could be available to species of all body sizes, but with different-sized species using different proportions of that available. However, energetic constraints can only apply if appropriated energy does reflect that available; if it does not, energy cannot be a constraint. If, on the other hand, energy available and energy used are related, the effects of the former will still depend on the number of species among which energy is allocated.

In summary, it is impossible to say anything about how energy availability might constrain any aspect of the abundance–body size relationship without knowing anything about the allometry of energy availability. However, even given that information, it is not clear that predictions about the shape of the abundance–body size relationship can be made from any pattern of energy availability alone.

Ultimately, the amount of life that can be sustained in a system must be limited by the amount of energy entering it. Very crudely, energy can be thought of as translating into biomass, which in turn translates into a certain number of individual animals (Lawton, 1990; Blackburn and Gaston, 1996b). At one level, therefore, energy can be considered to limit abundance only if all those individuals belong to a single species. As soon as there are two species, abundance is limited by competition between them for the available energy. The same argument applies at any level at which the energy available is not infinite; for example, if the energy available to species of a given body mass is limited, if the energy available to a trophic group is limited, and so on. While it could still be argued that it is energy that limits abundance, rather than the interspecific interaction, because abundance is ultimately set by the energy that is available after the interaction has been accounted for, this equally applies whatever the observed patterns of energy use and abundance. Therefore, none of the evidence currently used to support the idea of energetic limitation really does so. The more interesting question may be how energy translates through biomass and number of individuals into a certain number of species (Lawton, 1990; Blackburn and Gaston, 1996b).

C. The Latitudinal Gradient Mechanism

1. Description

Both body sizes and abundances (measured either as population sizes or densities) often show latitudinal gradients across species in their mean expression (e.g. Zeveloff and Boyce, 1988; Currie and Fritz, 1993; Cushman *et al.,* 1993; Barlow, 1994; Taylor and Gotelli, 1994; Hawkins and Lawton, 1995; Kaspari and Vargo, 1995; Poulin, 1995; Blackburn and Gaston, 1996c; Gaston and Blackburn, 1996a). The negative relationship between abundance and body size could perhaps arise as a result of compiling data from a range of latitudes; at any single latitude there might be no abundance–body size relationship.

2. Evaluation

In fact, it is extremely unlikely that the latitudinal gradient mechanism could cause the negative abundance–body size relationship. Where latitudinal gradients in abundance and body size exist, they run such that equatorial species tend, on average, to have lower population sizes, densities and body sizes (although see Currie and Fritz, 1993) than do species residing at higher latitudes. These patterns alone would generate a positive abundance-body size relationship.

On the other hand, the existence of latitudinal gradients in abundance and body size implies that the global abundance–body size relationship for a taxon is not representative of the relationship at any given point in space. The global relationship suggests that mean densities should be higher where mean body size is lower, but the latitudinal gradients suggest that low average density and small average body size coincide. The conflict can be resolved if there is a correlation between the latitude that species inhabit and their position in the abundance–body size relationship. For example, if those species from low latitudes in Figure 1 tend to fall below the illustrated line with slope –1, and species from high latitudes fall above it, then mean density and body mass should both increase with latitude. The negative abundance–body size relationship often observed could then be the composite of a series of latitudinal slices of similar slope but different elevation.

There is some evidence for this explanation from mammals and from wildfowl. Peters and Raelson (1984) show that density–body mass relationships calculated across mammal species from extratropical biogeographic regions have consistently higher elevations for their fitted slopes than do relationships calculated across mammal species from tropical regions. The slope of the abundance–body mass relationship for those mammal species in Figure 1 considered tropical by John Damuth (personal communication to A. Purvis) has a higher elevation than the slope for those considered extra-tropical (intercept at 1 g $10^{3.78}$ versus $10^{4.34}$; ANCOVA, $F_{1,451} = 4.56$, $p = 0.033$). The statistical significance of this difference should be treated with extreme caution, however, because the slope values are also significantly different (– 0.71 versus – 0.84; test for homogeneity of slopes, $F_{1,450} = 5.86$, $p = 0.016$) although, in general, abundance–body mass relationships from tropical regions do not differ significantly in slope from those from temperate regions (ANOVA on the independent regression slope estimates analysed in Blackburn and Gaston (1997a), $F_{1,105} = 0.23$, $p = 0.62$).

Wildfowl species show no significant relationship between population size and body mass (Gaston and Blackburn, 1996a). However, for a given body mass, wildfowl species living further from the equator have larger population sizes (Figure 4; the same is true for crude estimates of population density). In fact, the interspecific population size–body mass relationship becomes weakly, but significantly, negative if the latitude at which species breed is controlled for (Gaston and Blackburn, 1996a). Further, wildfowl show positive relationships between population size and latitudinal mid-point of breeding range, and body mass and latitudinal mid-point of breeding range, within taxa (Gaston and Blackburn, 1996a); in other words, among closely related taxa, those living closer to the equator have lower population sizes and body sizes. All the above patterns would be expected if extra-tropical taxa were consistently larger bodied and more abundant than their tropical relatives.

The different elevations of interspecific abundance–body size relationships in tropical and extra-tropical regions have two obvious consequences. First, if abundances are energy limited, higher abundances would be expected in the tropics, where levels of energy input are higher (and less variable?). Of course, species richness is also higher in the tropics, suggesting that there should be an interaction between abundance and species richness; this possibility is discussed at greater length elsewhere (Blackburn and Gaston, 1996b,c). Secondly, it suggests one reason why interspecific abundance–body size relationships compiled from the literature might tend to omit small-bodied rare species, as they have been suggested to do (Lawton, 1989). Such omission would arise if a higher proportion of small-bodied rare species are tropical, but the relationships include either higher proportions of non-tropical than tropical species, or a disproportionate number of large-bodied tropical species. In this last respect, it is interesting that the one study simultaneously to examine interspecific latitudinal patterns in body mass and density using data compiled from literature sources found positive latitudinal gradients in density, but negative gradients in body mass across a variety of taxa (Currie and Fritz, 1993). The body mass gradients run counter to most other studies (e.g. Blackburn and Gaston, 1996c, and references therein), and strongly suggest that tropical studies focus on the larger-bodied species. Where these latitudinal gradients have

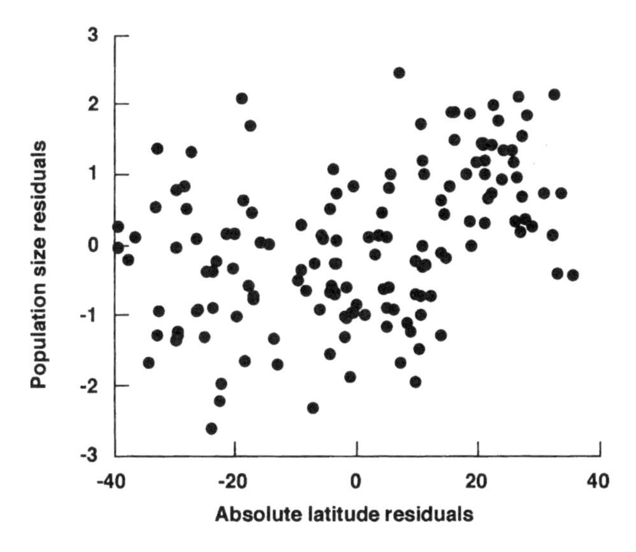

Fig. 4. The relationship between population size ($\log_{10}$ number of individuals) and the absolute latitudinal mid-point of a species' geographic range for the wildfowl of the world, controlling for ($\log_{10}$ transformed) species body mass (both variables are residuals from the simple regression on $\log_{10}$-transformed body mass, with mass as the independent variable) ($r^2 = 0.17$, $n = 147$, $p < 0.0001$). Species which live further from the equator have higher total population sizes for a given body mass.

been tested using all species in the taxon in question, they are positive for both body mass and abundance (Gaston and Blackburn, 1996a).

D. The Concatenation Mechanism

1. Description

The concatenation mechanism suggests that any observed abundance–body size relationship is a simple consequence of random sampling from the frequency distributions of abundance and body size for the taxon in question (Blackburn *et al.,* 1993b). It is related to the sampling models of Blackburn *et al.* (1990) and Currie (1993), but extends these treatments to produce quantitative predictions for the slope expected from random samples given certain explicit assumptions about the underlying distributions of species. While this mechanism is easily stated, its derivation is more complicated. We confine ourselves to the essentials here; a fuller treatment can be found in Blackburn *et al.* (1993b).

The concatenation mechanism assumes that the potential abundances of species in a taxon can be reasonably well described by a log-normal frequency distribution (see Preston, 1948, 1962; Sugihara, 1980; Tokeshi, 1990; Nee *et al.,* 1991a; Gregory, 1994). In a random sample from a log-normal distribution, the range R (scaled to s, the standard deviation of the distribution) of abundances expected is proportional to the logarithm of the sample size, when the sample size is of the order of 10 to 1000 species (Tippett, 1925; Gumbel, 1947). We know that there are fewer large-bodied than small-bodied species in most, if not all, taxa (e.g. May, 1978; Blackburn and Gaston, 1994; Brown, 1995). Therefore, the higher maximum abundances of small-bodied species may arise simply because small-bodied species represent a larger sample from the underlying abundance distribution. A negative relationship between maximum abundance and body size would be expected from random sampling alone.

The simple random sampling model above alone is not sufficient to explain a general negative abundance–body size relationship, because it also predicts that the decrease in maximum abundance with increasing body size should be reflected by an increase in minimum abundance. This should result in a roughly triangular abundance–body size relationship (with the triangle here like an arrowhead pointing towards the largest species). An approximately linear negative relationship would be expected, though, if the random sampling model was coupled with a minimum viable abundance constraint. The constraint would restrict the part of the frequency distribution from which random samples could be drawn for species of different body sizes. To work, it requires that small-bodied species have higher minimum viable abundances (either density or population size) than large-bodied species, but makes no assumption about the magnitude of the difference. This is a point to which we return later.

2. Evaluation

There is some evidence for this mechanism, in that the upper bound slopes of real assemblages tend to fit the predictions of the random sampling model with a minimum viable abundance constraint (Blackburn *et al.,* 1993a,b). It can also explain the decrease in abundance observed at the smallest body sizes in some taxa (Brown and Maurer, 1987; Brown, 1995; Silva and Downing, 1995), because the maximum abundance in a taxon should mirror the frequency distribution of body sizes, which is normally right-skewed under logarithmic transformation (log-right skewed; Blackburn and Gaston, 1994).

However, the theoretical requirements of the concatenation mechanism suggest that alone it may not be sufficient to explain the observed abundance–body size patterns. Specifically, the minimum viable abundance constraint requires that abundances are random samples from a potential log-normal distribution and not from an actual distribution. It is easy to see why, if one imagines random abundance samples for small- and large-bodied species (Figure 5). There will be few large-bodied species, so their abundances should on average be sampled from under the peak of the frequency distribution. The abundances randomly sampled for the greater number of small-bodied species should show a greater range but the minimum viable population constraint dictates that these abundances should not be less than those of the large-bodied species. In other words, for the sampling mechanism to work, no species actually has an abundance from the lower tail of the abundance distribution, which can then only be a potential, not the actual, frequency distribution. The mechanism implies that there are lots of low abundances that species potentially can attain, but actually never do. While this is certainly true, it should also be true for high abundances; it is difficult to argue for a log-normal shape for such an imaginary distribution.

In their original paper, Blackburn *et al.* (1993b) argued that the minimum viable abundance constraint could cause the entire frequency distribution to be shifted up at small body sizes, so that the abundances of small species are effectively sampled from a distribution with the same variance but different mean. However, this formulation does not avoid the above problems; abundances are still never sampled from the lower tail of the abundance distribution for all species in the assemblage. Alternatively, to reverse the argument, the abundances of large species are constrained always to come from the lower tail of the real abundance distribution, because the abundances of small species will always tend to be higher; such a situation can hardly be described as "random sampling". The above arguments suggest that the concatenation mechanism is not a sufficient explanation for negative abundance–body size relationships, although sampling problems undoubtedly can contribute to patterns caused by other mechanisms (Blackburn *et al.,* 1990; Currie, 1993).

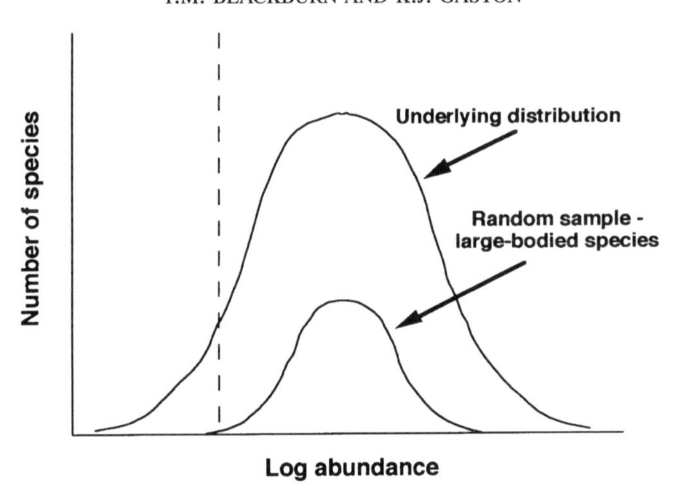

Fig. 5. A hypothetical species abundance distribution. Under the concatenation mechanism, samples are taken at random from the underlying distribution. The second distribution shows one such random sample, assumed here to represent large-bodied species. The minimum population size constraint means that no smaller-bodied species can have an abundance less than the lowest exhibited by large-bodied species; no species can have an abundance that falls to the left of the dashed line. Therefore, the random samples are not true samples of the actual underlying distribution illustrated, but some other distribution. It is difficult to predict a *priori* what shape this other distribution might be. Note that this argument still applies if the illustrated underlying distribution is truncated at the dashed line (as many real distributions are) because there are still parts of the underlying distribution that are excluded from some of the "random" samples.

E. Interspecific Competition

1. Description

Somewhat surprisingly, given the preoccupations of a large section of the ecological literature, interspecific competition has never, to our knowledge, been invoked to explain the full interspecific abundance–body size relationship. Nonetheless, competition does seem able to explain some of the peculiarities of the patterns observed within taxa. Within taxa, positive abundance–body size relationships are common at low taxonomic levels (e.g. across species within a genus), but negative relationships predominate at higher levels (e.g. across superfamilies within orders; Nee *et al.,* 1991b; Blackburn *et al.,* 1994; Gregory, 1995). Further positive relationships across species within tribes tend to be associated with tribes of greater evolutionary age (Cotgreave and Harvey, 1991; Nee *et al.,* 1991b; Blackburn *et al.,* 1994; Cotgreave, 1994). The association between tribe age and patterns in abundance is especially significant, because abundances change orders of magnitude more rapidly than lineages evolve.

The idea that competition could affect the within-taxon abundance–body size relationship is owed mainly to the work of Nee and Cotgreave (e.g. Cotgreave and Harvey, 1991, 1994; Nee *et al.,* 1991b, 1992; Cotgreave, 1994; Cotgreave and Stockley, 1994). They reasoned that interspecific competition should be most intense between members of an ecological guild, which are often closely related. If large body size is an advantage in interspecific competition, leading to higher levels of abundance in larger species when competition is important, then positive relationships between abundance and body size would be expected in comparisons between species in the same guild (or at least between species in the same taxon, which are easier to define). Further, if evolutionarily isolated tribes also tend to form complete guilds, then an association between evolutionary isolation and the slope of the abundance–body size relationship would also be expected (Nee *et al.,* 1992; Cotgreave, 1994). An oft-quoted example is the woodpecker tribe (Picidae), which is both phylogenetically isolated and arguably forms a complete guild (e.g. all its species have similar nesting and feeding habits not shared by any other species; Cotgreave, 1994). The abundance–body size relationship in this tribe is generally positive (Figure 6) (Nee *et al.,* 1991b; Blackburn *et al.,* 1994).

2. Evaluation

The chain of reasoning connecting the pattern of abundance–body size relationships within-taxa to competition is long and potentially fragile. However, there are two analyses that support it.

First, Bock *et al.* (1992) studied the abundances of insectivorous birds in a riparian community in Arizona, USA, before and after the erection of nest boxes in the experimental area. The presence of nest boxes caused some species to increase in abundance and others to decrease. Bock *et al.* argued that those species that increased did so because they were freed from competition for nest sites, while those that decreased did so because of increased competition for food with the species whose abundances increased. In other words, competition was normally more important as a force limiting abundance for those species that increased in abundance after the experimental treatment. Cotgreave (1994) examined the relationship between abundance and body size within tribes in the data published by Bock *et al.* As in other studies the date of origin of the tribe was positively correlated with the slope of the relationship within a tribe. However, so too was the percentage change in abundance following the experimental treatment, and the proportion of cavity-nesting species; tribes with positive abundance–body size relationships tended to increase in abundance and be hole nesters, and those with negative relationships all decreased, and tended not to be hole nesters. This is as would be expected if high competition for nest sites led to positive relationships. Nevertheless, the study of Bock *et al.* only exam-

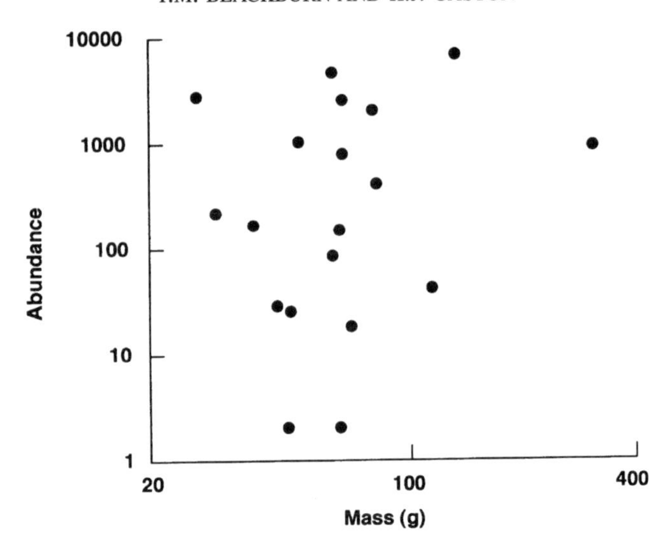

Fig. 6. The relationship between body mass (g) and abundance (total number of individuals recorded on all US and Canadian Fish and Wildlife Service Breeding Bird Survey routes for the year 1977; data from Robbins *et al.*, 1986) in North American woodpeckers. The relationship is positive, although not significantly so ($n = 19$).

ines the effect of competition for nest sites. It is unclear whether taxa with positive abundance–body size relationships in this assemblage experience generally higher levels of competition.

Secondly, Robinson and Terborgh (1995) examined interspecific competition in territorial bird species in Amazonian rain forest. For 20 congeneric species pairs with non-overlapping territories, they used song playback experiments to demonstrate that the larger-bodied species in the pair responded aggressively to the song of the smaller-bodied species, but that the smaller-bodied species did not respond aggressively to the song of the larger-bodied species. Robinson and Terborgh argued that this resulted in the larger-bodied species occupying more productive habitats, and therefore being able to attain higher densities than their smaller congeners. The expected corollary, that small-bodied species should be commoner in species pairs where competition is not important, remains untested.

More indirect evidence for the effect of competition comes from studies of niche overlap between closely related species in animal assemblages. The argument is that competition will be strongest where niche overlap is greatest (Cotgreave and Harvey, 1994; Cotgreave and Stockley, 1994; Cotgreave, 1995). Positive abundance–body size relationships should therefore be found in taxa with high niche overlap. This does indeed seem to be the case in small mammal assemblages (Cotgreave and Stockley, 1994) and for Australian birds (Cotgreave, 1995). The latter example is interesting, because high niche overlap is associated with positive within-tribe abundance–body size relationships, but older tribe age is not. If tribe age is a surrogate for level of competition in some assemblages, we would expect the effect of competition to be expressed even in the absence of an age effect.

As an aside, the interspecific competition hypothesis suggests a reason why large body size might be favoured in some, but not all, taxa. That could potentially explain both the phenomenon known as Cope's Rule (Cope, 1887, 1896; cited in Stanley, 1973)—the tendency for body size to increase in taxa through evolutionary time—and why this phenomenon is only observed in some taxa. However, firm evidence for such a link could be hard to establish.

If interspecific competition does have the potential to explain abundance–body size relationships at low taxonomic levels, it is pertinent to ask whether it also has the potential to explain higher level patterns. The short answer would seem to be that it is unlikely. Competition is most probable between members of ecological guilds (as defined above). However, there is already evidence that guilds tend to occupy different parts of abundance–body size space (e.g. Peters and Wassenberg, 1983; Peters and Raelson, 1984; Juanes, 1986; Damuth, 1987; Silva and Downing, 1995). The relative positions of different guilds are more likely to be determined by the ecological requirements of guild membership, rather than competition between members of different guilds. If positive abundance–body size relationships are the result of competition, but competition is most important at low taxonomic levels (e.g. within genera), that would explain both the paucity of positive relationships at higher levels (e.g. across orders), and also why some positive interspecific relationships do exist (Blackburn and Gaston, 1997).

F. The Differential Extinction Mechanism

1. Description

The second mechanism proposed to explain within-taxon abundance–body size relationships invokes patterns of extinction as the cause of positive relationships across species within older tribes (Blackburn *et al.,* 1994). Large-bodied species tend, at least in some cases, to be more susceptible to extinction within taxa than small-bodied species (Gaston and Blackburn, 1995). Therefore, tribes may differentially lose large-bodied rare species to extinction, but not small-bodied species of equivalent abundance. The older the tribe, the greater is the probability that large-bodied, rare species in it will have become extinct. Consequently, ancient tribes may consist only of common large-bodied species but both common and rare small-bodied species, causing the within-tribe relationship between abundance and body size to be positive. This would result in a tendency for positive abundance–body size relationships in older tribes.

2. Evaluation

At present, there is little evidence for this mechanism. It assumes that newly evolved taxa should contain both small and large, common and rare species, and so, on average, show no relationship between abundance and body size. This fits with the observation that there is no strong tendency for positive or negative relationships to predominate at the lowest taxonomic levels (e.g. across species within genera; Nee *et al.*, 1991b; Blackburn *et al.*, 1994). However, while the differential extinction hypothesis is plausible given what we know about body size-dependent extinction within taxa (Gaston and Blackburn, 1995), these observed extinction rates are not independent of abundance. Hence, it is not clear that small-bodied species should be more extinction-proof than large-bodied species of equivalent low abundance; they may be more extinction-proof because they tend to have higher abundances. Further, it is inadequate as a sole explanation for positive within-taxon abundance–body size relationships, because additional mechanisms to drive the differential extinction rates are still required.

IV. DISCUSSION

No single mechanism that has been suggested to explain the interspecific abundance–body size relationship can be considered adequate. In particular, the continued popularity and discussion of the energetic constraint mechanism is entirely unwarranted given that it is founded on an assumption for which there is no evidence, and that is almost certainly incorrect. That is not to say that, at the most fundamental level, energy does not determine the amount of life that can be sustained on the planet. Rather, there is as yet no plausible mechanism to explain how that energy is divided among species to give the relationship between abundance and body size that we observe in real data. Any mechanism that can account for abundance–body size relationships will also have to explain the latitudinal component to the relationship, and to incorporate different effects at different taxonomic levels. The problem that census area affects observed relationships to a degree that is unknown will continue to bedevil interpretation of evidence in support of particular hypotheses.

Since abundance, body mass and biomass are clearly all linked, to what extent can patterns in the distributions of abundance, body mass, and the relationship between the two, be derived from the frequency distribution of biomasses? Given a few simple assumptions, such derivations are relatively straightforward. Specifically, if it is assumed that:

(1) the amount of biomass that can be maintained in a region depends on, and is proportional to, the amount of energy available there;
(2) this biomass is divided among species so as to produce a log-normal biomass frequency distribution;

(3) there is a minimum viable abundance (population size or density) below which populations quickly become extinct;

then the result is an assemblage of species, each consisting of a certain amount of biomass (Lawton, 1990). This biomass is divided into a number of individuals of a given size. Size could be determined for each species in a number of ways, for example, to maximise lifetime reproductive fitness in the manner suggested by Kozlowski and Weiner (1996). Average individual size and species biomass set the species abundance. The actual amount of biomass a species has sets its elevation on the abundance–body mass relationship; for a given size, species with less biomass have fewer individuals but, for a given biomass, species with smaller individuals have more of them. Biomass cannot be divided too finely, because the minimum viable abundance constraint sets a lower limit to the amount that a species requires (Hutchinson, 1959; Lawton, 1990). Nevertheless, this minimum biomass can still be divided into fewer large-bodied or more small-bodied individuals. The result is a negative abundance–body size relationship, in that the maximum achievable abundance will decline with increasing body size. If the lower bound to abundance is assumed to be a declining function of body size (as in the concatenation mechanism, see above), then the abundance–body size relationship will tend to become more linear and less triangular, depending fundamentally on how steep this bound is.

How valid are the three assumptions of this model? First, a positive relationship between biomass and available energy seems reasonable in many circumstances. Biomass and primary productivity have been shown to be broadly positively correlated across both terrestrial and aquatic systems, at least for herbivores (McNaughton *et al.,* 1989; Cyr and Pace, 1993).

Secondly, the species–biomass distribution does seem likely to be roughly log-normal. For example, Figure 7 gives the biomass distribution for three different assemblages of birds, where biomass was obtained in each case by multiplying the number of individuals of each species in the assemblage by its mean body mass. These biomass estimates are likely to be crude, ignoring as they do variances in the abundances and body masses of the individual species. Nevertheless they strongly suggest that a log-normal distribution is a good first approximation.

The key process in the above model is how total biomass is divided between species to yield the biomass frequency distribution (see also Lawton, 1990). Several models have been suggested for producing the log-normal abundance distribution (reviewed by Tokeshi, 1990), which could reasonably also work for biomass. One of the more attractive is Sugihara's (1980) sequential niche breakage (sequential "broken stick") model. This posits that successive species to invade an environment randomly annex a proportion of the resources already utilised by one of the species present. This process produces an approximately log-normal distribution of resources held by the final assemblage of species (in fact, communities simulated under this model tend to have abundance distributions that are left-skewed (Nee *et al.,* 1991a), as are the biomass distributions in Figure 7). Exactly how those resources translate into abundances has been contentious (Harvey and Godfray, 1987; Sugihara, 1989; Pagel *et al.,* 1991), but the translation between resource and biomass should be less so, being more direct. One interesting point is that, in the real world, species invading the assemblage late on are more likely to subsume part of the resources of an ecologically similar species. At this level, it is easy to see how competition between ecologically similar species for part of a piece of the resource spectrum may give rise to the relationships observed within taxa at low taxonomic levels (see section III.E).

The third assumption concerns the lower bound to the abundance–body size relationship. That there is an abundance below which a species cannot persist is necessarily true. However, consideration of the probable form of the lower bound opens up a complex set of issues, which are listed below.

(1) Some species in an assemblage may not have viable abundances (particularly in local assemblages).
(2) Viability cannot be determined without reference to a time frame. Extinction is inevitable, so even huge populations have a finite probability of becoming extinct over any time period.
(3) For a given time frame and risk of extinction, minimum viable abundances may or may not change systematically with body size. While it is widely assumed that, for a given abundance, species of larger body size are at greater risk of extinction, the evidence is equivocal, and reasons can be suggested why it is not true (see Gaston and Blackburn, 1996b).
(4) Abundance–body size relationships tend predominantly to be founded on densities, but viable abundances concern population sizes and it is not clear what the relationship between the two actually is (Lawton, 1990).

The first two issues complicate analyses of patterns in minimum viable abundance (e.g. Silva and Downing, 1994), because they imply that comparable viabilities are very difficult to obtain. It is the third and fourth issues that, in practice, should determine the form of the lower bound to the abundance–body size relationship. These issues will also affect the frequency distributions of body mass and abundance, the exact shapes of which will be determined in part by the optimum trade-off between body size and population size in those species near the left-hand end of the biomass distribution (e.g. Figure 7). However, the form of the relationship between body size and minimum viable population size is currently controversial (e.g. Pimm *et al.,* 1988; Tracy and George, 1992, 1993; Diamond and Pimm, 1993; Haila and Hanski, 1993). Also, while the relationship between minimum viable population size and density would benefit from explicit study, that will not be

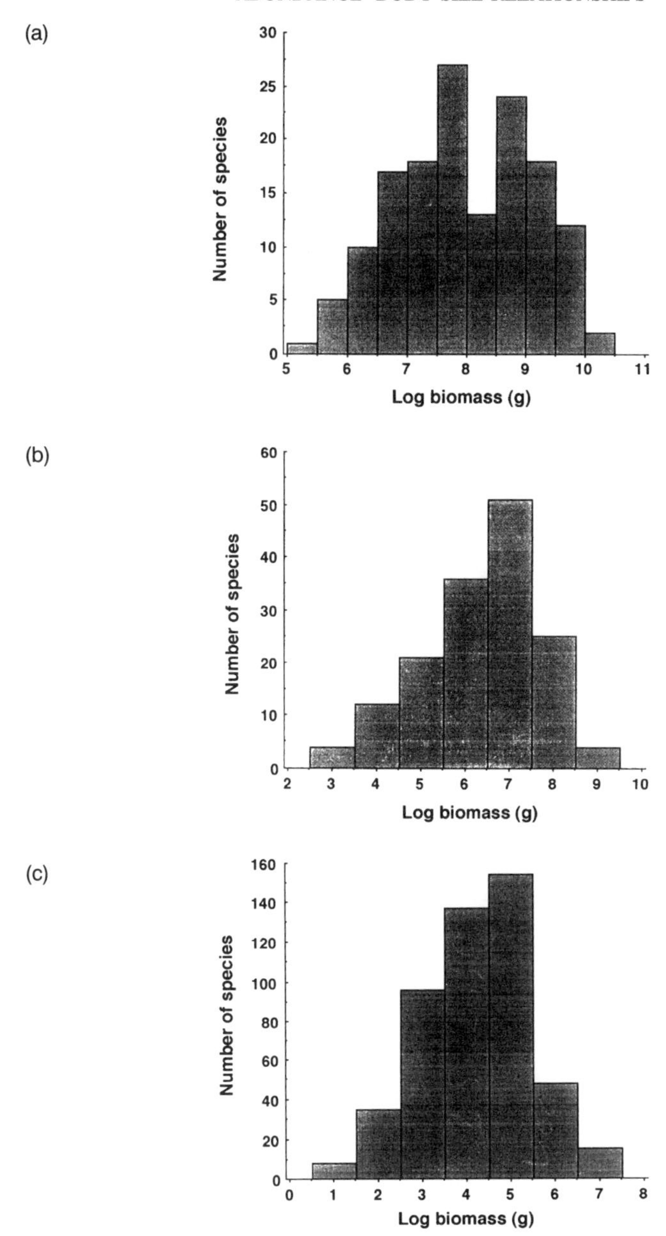

Fig. 7. The frequency distribution of $\log_{10}$ biomass (g) for (a) wildfowl of the world (data as used in Gaston and Blackburn, 1996a); (b) British breeding birds (data as used in Blackburn *et al.*, 1996); and (c) birds on North American Breeding Bird Survey census routes in 1977 (data on numbers of individuals from Robbins *et al.* (1986, Table 1); data on body masses were kindly supplied by Brian Maurer).

straightforward given difficulties arising from the definitions of both "viable" and "density".

In conclusion, none of the standard hypotheses for the form of the abundance–body size relationship are satisfactory. In retrospect, this is perhaps unsurprising; explanations that focus on the interspecific negative relationship will almost inevitably fail, because they ignore the fact that abundance and body size show a variety of associations, and no overall relationship, within taxa. In particular, it is these phylogenetic patterns that imply to us that there is unlikely to be a general causal link between the variables. For this reason, we suggest that negative interspecific abundance–body size relationships do not need separate explanations beyond those needed to explain, first, why there is an approximately log-normal biomass distribution and, secondly, the form of the relationship between minimum viable abundance and body size; patterns among closely related taxa in ecological guilds may be fine-tuned by interspecific competition. However, given that explanations for the abundance distribution have generated considerable debate with little clear resolution (e.g. Preston, 1948, 1962; May, 1975; Sugihara, 1980, 1989; Gray, 1987; Harvey and Godfray, 1987; Tokeshi, 1990; Nee *et al.,* 1991a; Pagel *et al.,* 1991; Gregory, 1994), and that studies of minimum viable abundances need to overcome serious analytical problems, our suggestion may be simplifying only in terms of the number of explanations that are required. This particular application of Occam's razor could prove an unkind cut.

ACKNOWLEDGMENTS

K.J.G. is a Royal Society University Research Fellow. T.M.B. was funded by NERC grant GST/03/1211, and by the core grant to the NERC Centre for Population Biology. We thank John Lawton, Dave Raffaelli and an anonymous referee for helpful comments on the manuscript.

REFERENCES

Barlow, N.D. (1994). Size distributions of butterfly species and the effect of latitude on species sizes. *Oikos* **71**, 326–332.

Basset, Y. and Kitching, R.L. (1991). Species number, species abundance and body length of arboreal arthropods associated with an Australian rainforest tree. *Ecol. Ent.* **16**, 391–402.

Blackburn, T.M. and Gaston, K.J. (1994). Animal body size distributions: Patterns, mechanisms and implications. *Trends Ecol. Evol.* **9**, 471–474.

Blackburn, T.M. and Gaston, K.J. (1996a). Abundance–body size relationships: The area you census tells you more. *Oikos* **75**, 303–309.

Blackburn, T.M. and Gaston, K.J. (1996b). A sideways look at patterns in species richness, or why there are so few species outside the tropics. *Biodiv. Lett.* **3**, 44–53.

Blackburn, T.M. and Gatson, K.J. (1996c). Spatial patterns in the body sizes of bird species in the New World. *Oikos* **77**, 436–446.

Blackburn, T.M. and Gaston, K.J. (1997a). A critical assessment of the form of the interspecific relationship between abundance and body size in animals. *J. Anim. Ecol.* **66**, 233–249.

Blackburn T.M. and Gaston, K.J. (1997b). Who is rare? Artefacts and complexities in rarity determination. In: *The Biology of Rarity* (Ed. by W.E. Kunin and K.J. Gaston), pp. 48–60. Chapman and Hall, London.

Blackburn, T.M. and Lawton, J.H. (1994). Population abundance and body size in animal assemblages. *Phil. Trans. Roy. Soc. B* **343**, 33–39.

Blackburn, T.M., Harvey, P.H. and Pagel, M.D. (1990). Species number, population density and body size in natural communities. *J. Anim. Ecol.* **59**, 335–346.

Blackburn, T.M., Brown, V.K., Doube, B.M., Greenwood, J.J.D., Lawton, J.H. and Stork, N.E. (1993a). The relationship between body size and abundance in natural animal assemblages. *J. Anim. Ecol.* **62**, 519–528.

Blackburn, T.M., Lawton, J.H. and Pimm, S.L. (1993b). Non-metabolic explanations for the relationship between body size and animal abundance. *J. Anim. Ecol.* **62**, 694–702.

Blackburn, T.M., Gates, S., Lawton, J.H. and Greenwood, J.J.D. (1994). Relations between body size, abundance and taxonomy of birds wintering in Britain and Ireland. *Phil. Trans. Roy. Soc. B* **343**, 135–144.

Blackburn, T.M., Lawton, J.H. and Gregory, R.D. (1996). Relationships between abundances and life histories of British birds. *J. Anim. Ecol.* **65**, 52–62.

Bock, C.E., Cruz, A., Grant, M.C., Aid, C.S. and Strong, T.R. (1992). Field experimental evidence for diffuse competition among southwestern riparian birds. *Am. Nat.* **140**, 815–828.

Brown, J.H. (1995). *Macroecology.* University of Chicago Press, Chicago.

Brown, J.H. and Maurer, B.A. (1986). Body size, ecological dominance and Cope's Rule. *Nature* **324**, 248–250.

Brown, J.H. and Maurer, B.A. (1987) Evolution of species assemblages: Effects of energetic constraints and species dynamics on the diversification of the American avifauna. *Am. Nat.* **130**, 1–17.

Brown, J.H. and Maurer, B.A. (1989). Macroecology: The division of food and space among species on continents. *Science* **243**, 1145–1150.

Cope, E.D. (1887). *The Origin of the Fittest.* D. Appleton and Co., New York.

Cope, E.D. (1896). *The Primary Factors of Organic Evolution.* Open Court Publ. Co., Chicago.

Cotgreave, P. (1993). The relationship between body size and population abundance in animals. *Trends Ecol. Evol.* **8**, 244–248.

Cotgreave, P. (1994). The relationship between body size and abundance in a bird community: The effects of phylogeny and competition. *Proc. R. Soc. B* **256**, 147–149.

Cotgreave, P. (1995). Population density, body mass and niche overlap in Australian birds. *Funct. Ecol.* **9**, 285–289.

Cotgreave, P. and Harvey, P.H. (1991). Bird community structure. *Nature* **353**, 123.

Cotgreave, P. and Harvey, P.H. (1994). Phylogeny and the relationship between body size and abundance in bird communities. *Funct. Ecol.* **8**, 219–228.

Cotgreave, P. and Stockley, P. (1994). Body size, insectivory and abundance in assemblages of small mammals. *Oikos* **71**, 89–96.

Currie, D.J. (1993). What shape is the relationship between body size and population density? *Oikos* **66**, 353–358.

Currie, D.J. and Fritz, J.T. (1993). Global patterns of animal abundance and species energy use. *Oikos* **67**, 56–68.

Cushman, J.H., Lawton, J.H. and Manly, B.F.J. (1993). Latitudinal patterns in European ant assemblages: Variation in species richness and body size. *Oecologica* **95**, 30–37.

Cyr, H. and Pace, M.L. (1993). Magnitude and patterns of herbivory in aquatic and terrestrial ecosystems. *Nature* **361**, 148–150.

Damuth, J. (1981). Population density and body size in mammals. *Nature* **290**, 699–700.

Damuth, J. (1987). Interspecific allometry of population density in mammals and other animals: The independence of body mass and population energy use. *Biol. J. Linn. Soc.* **31**, 193–246.

Diamond, J. and Pimm, S.L. (1993). Survival times of bird populations: A reply. *Am. Nat.* **142**, 1030–1035.

Elgar, M.A. and Harvey, P.H. (1987). Basal metabolic rates in mammals: Allometry, phylogeny and ecology. *Funct. Ecol.* **1**, 25–36.

Gaston, K.J. (1988). Patterns in the local and regional dynamics of moth populations. *Oikos* **53**, 49–57.

Gaston, K.J. (1994). *Rarity.* Chapman and Hall, London.

Gaston, K.J. and Blackburn, T.M. (1995). Birds, body size and the threat of extinction. *Phil. Trans. R. Soc. Lond. B* **347**, 205–212.

Gaston, K.J. and Blackburn, T.M. (1996a). Global scale macroecology: Interactions between population size, geographic range size and body size in the Anseriformes. *J. Anim. Ecol.* **65**, 701–714.

Gaston, K.J. and Blackburn, T.M. (1996b). Some conservation implications of geographic range size–body size relationships. *Conserv. Biol.* **10**, 638–646.

Gray, J.S. (1987). Species-abundance patterns. In: *The Organisation of Communities: Past and Present* (Ed. by J.H.R. Gee and P.S. Giller), pp. 53–67. Blackwell Scientific Publications, Oxford.

Greenwood, J.J.D., Gregory, R.D., Harris, S., Morris, P.A. and Yalden, D.W. (1996). Relations between abundance, body size and species number in British birds and mammals. *Phil. Trans. R. Soc. B* **351**, 265–278.

Gregory, R.D. (1994). Species abundance patterns of British birds. *Proc. R. Soc. Lond. B* **257**, 299–301.

Gregory, R.D. (1995). Phylogeny and relations among abundance, geographical range and body size of British breeding birds. *Phil. Trans. R. Soc. B* **349**, 345–351.

Gregory, R.D. and Blackburn, T.M. (1995). Abundance and body size in British birds: Reconciling regional and ecological densities. *Oikos* **72**, 151–154.

Gumbel, E.J. (1947). The distribution of the range. *Ann. Mathemat. Stat.* **18**, 384–412.

Haila, Y. (1988). Calculating and miscalculating density: The role of habitat geometry. *Ornis Scand.* **19**, 88–92.

Haila, Y. and Hanski, I.K. (1993). Birds breeding on small islands and extinction risks. *Am. Nat.* **142**, 1025–1029.

Harvey, P.H. (1996). Phylogenies for ecologists. *J. Anim. Ecol.* **65**, 255–263.

Harvey, P.H. and Godfray, H.C.J. (1987). How species divide resources. *Am. Nat.* **129**, 318–320.

Harvey, P.H. and Lawton, J.H. (1986). Patterns in three dimensions. *Nature* **324**, 212.

Hawkins, B.A. and Lawton, J.H. (1995). Latitudinal gradients in butterfly body sizes: Is there a general pattern? *Oecologia* **102**, 31–36.

Hutchinson, G.E. (1959). Homage to Santa Rosalia or why are there so many kinds of animals? *Am. Nat.* **93**, 145–159.

Juanes, F. (1986). Population density and body size in birds. *Am. Nat.* **128**, 921–929.

Kaspari, M. and Vargo, E.L. (1995). Colony size as a buffer against seasonality: Bergmann's rule in social insects. *Am. Nat.* **145**, 610–632.

Kleiber, M. (1962). *The Fire of Life.* Wiley, New York.

Kozlowski, J. and Weiner, J. (1996). Interspecific allometries are byproducts of body size optimization. *Am. Nat.* (in press).

Lawton, J.H. (1989). What is the relationship between population density and body size in animals? *Oikos* **55**, 429–434.

Lawton, J.H. (1990). Species richness and population dynamics of animal assemblages. Patterns in body-size: Abundance space. *Phil. Trans. Roy. Soc. B* **330**, 283–291.

McNaughton, S.J., Oesterheld, M., Frank, D.A. and Williams, K.J. (1989). Ecosystem-level patterns of primary productivity and herbivory in terrestrial habitats. *Nature* **341**, 142–144.

Marquet, P.A., Navarette, S.A. and Castilla, J.C. (1990). Scaling population density to body size in rocky intertidal communities. *Science* **250**, 1125–1127.

Marquet, P.A., Navarette, S.A. and Castilla, J.C. (1995). Body size, population density, and the energetic equivalence rule. *J. Anim. Ecol.* **64**, 325–332.

May, R.M. (1975). Patterns of species abundance and diversity. In: *Ecology and Evolution of Communities* (Ed. by M.L. Cody and J.M. Diamond), pp. 81–120. Harvard University Press, Cambridge, MA.

May, R.M. (1978). The dynamics and diversity of insect faunas. In: *Diversity of Insect Faunas* (Ed. by L.A. Mound and N. Waloff), pp. 188–204. Blackwell Scientific Publications, Oxford.

Medel, R.G., Bozinovic, F. and Novoa, F.F. (1995). The mass exponent in population energy use: The fallacy of averages reconsidered. *Am. Nat.* **145**, 155–162.

Morse, D.R., Stork, N.E. and Lawton, J.H. (1988). Species number, species abundance and body length relationships of arboreal beetles in Bornean lowland rain forest trees. *Ecol. Entomol.* **13**, 25–37.

Nagy, K.A. (1987). Field metabolic rate and food requirement scaling in mammals and birds. *Ecol. Mon.* **57**, 111–128.

Nee, S., Harvey, P.H. and Cotgreave, P. (1992). Population persistence and the natural relationships between body size and abundance. In: *Conservation of Biodiversity for Sustainable Development* (Ed. by O.T. Sandlund, K. Hindar and A.H.D. Brown), pp. 124–136. Scandinavian University Press, Oslo.

Nee, S., Harvey, P.H. and May, R.M. (1991a). Lifting the veil on abundance patterns. *Proc. R. Soc. Lond. B* **243**, 161–163.

Nee, S., Read, A.F., Greenwood, J.J.D. and Harvey, P.H. (1991b). The relationship between abundance and body size in British birds. *Nature* **351**, 312–313.

Pagel, M.D., Harvey, P.H. and Godfray, H.C.J. (1991). Species abundance, biomass and resource use distributions. *Am. Nat.* **138**, 836–850.

Peters, R.H. (1983). *The Ecological Implications of Body Size.* Cambridge University Press, Cambridge.

Peters, R.H. and Raelson, J.V. (1984). Relationships between individual size and mammalian population density. *Am. Nat.* **124**, 498–517.

Peters, R.H. and Wassenberg, K. (1983). The effect of body size on animal abundance. *Oecologia* **60**, 89–96.

Pimm, S.L., Jones, H.L. and Diamond, J. (1988). On the risk of extinction. *Am. Nat.* **132**, 757–785.

Poulin, R. (1995). Evolutionary influences on body size in free-living and parasitic isopods. *Biol. J. Linn. Soc.* **54**, 231–244.

Preston, F.W. (1948). The commonness, and rarity, of species. *Ecology* **29**, 254–283.

Preston, F.W. (1962). The canonical distribution of commonness and rarity. *Ecology* **43**, 185–215, 410–432.

Reiss, M.J. (1989). *The Allometry of Growth and Reproduction.* Cambridge University Press, Cambridge.

Robbins, C.S., Bystrak, D. and Geissler, P.H. (1986). *The Breeding Bird Survey: Its First Fifteen Years, 1965–1979.* U.S. Department of the Interior Fish and Wildlife Service, Resource Publication 157, Washington, DC.

Robinson, S.K. and Terborgh, J. (1995). Interspecific aggression and habitat selection by Amazonian birds. *J. Anim. Ecol.* **64**, 1–11.

Schonewald-Cox, C., Azari, R. and Blume, S. (1991). Scale, variable density, and conservation planning for mammalian carnivores. *Cons. Biol.* **5**, 491–495.

Seber, G.A.F. (1982). *The Estimation of Animal Abundance and Related Parameters*, 2nd edn. Macmillan, New York.

Silva, M. and Downing, J.A. (1994). Allometric scaling of minimal mammal densities. *Conserv. Biol.* **8**, 732–743.

Silva, M. and Downing, J.A. (1995). The allometric scaling of density and body mass: A nonlinear relationship for terrestrial mammals. *Am. Nat.* **145**, 704–727.

Smallwood, K.S. and Schonewald, C. (1996). Scaling population density and spatial pattern for terrestrial mammalian carnivores. *Oecologia* **105**, 329–335.

Smallwood, K.S., Jones, G. and Schonewald, C. (1996). Spatial scaling of allometry for terrestrial, mammalian carnivores. *Oecologia* **107**, 588–594.

Solonen, T. (1994). Finnish bird fauna—species dynamics and adaptive constraints. *Ornis Fenn.* **71**, 81–94.

Stanley, S.M. (1973). An explanation for Cope's Rule. *Evolution* **27**, 1–26.

Strayer, D.L. (1994). Body size and abundance of benthic animals in Mirror Lake, New Hampshire. *Freshw. Biol.* **32**, 83–90.

Sugihara, G. (1980). Minimal community structure: An explanation of species abundance patterns. *Am. Nat.* **116**, 770–787.

Sugihara, G. (1989). How *do* species divide resources? *Am. Nat.* **133**, 458–463.

Taylor, C.M. and Gotelli, N.J. (1994). The macroecology of *Cyprinella*: Correlates of phylogeny, body size and geographical range. *Am. Nat.* **144**, 549–569.

Tippett, L.H.C. (1925). On the extreme individuals and the range of samples taken from a normal population. *Biometrika* **17**, 364–387.

Tokeshi, M. (1990). Niche apportionment or random assortment: Species abundance patterns revisited. *J. Anim. Ecol.* **59**, 1129–1146.

Tracy, C.R. and George, T.L. (1992). On the determinants of extinction. *Am. Nat.* **139**, 102–122.

Tracy, C.R. and George, T.L. (1993). Extinction probabilities for British Island birds: A reply. *Am. Nat.* **142**, 1036–1037.

Wagner, H.M. (1969). *Principles of Operation Research.* Prentice-Hall, Englewood Cliffs, NJ.

Welsh, A.H., Peterson, A.T. and Altmann, S.A. (1988). The fallacy of averages. *Am. Nat.* **132**, 277–288.

Zeveloff, S.I. and Boyce, M.S. (1988). Body size patterns in North American mammal faunas. In: *Evolution of Life Histories of Mammals* (Ed. by M.S. Boyce), pp. 123–146. Yale University Press, New Haven, CT.

Advances in Ecological Research Volumes 1–28

Cumulative List of Titles

Index